ETHER FLOWS

EXPLORING THE HIDDEN ENERGY THAT CONTROLS THE UNIVERSE

DANIEL N. SPATUCCI

© 2021 Daniel N. Spatucci
Published by Plasmic Studio LLC
www.plasmicstudio.com

Cover design and interior illustrations by Steve Spatucci

CONTENTS

INTRODUCTION

Absolutely true story: In 1975 I was newly married and living in San Francisco. It was late (about 2 or 3am), and I was playing guitar with the TV on in the background, when a show came on that was a recap of the day's news. As I remember it, there was national and local news, then some feature items, and finally a science news section. In this science section, they showed something that I don't think they were supposed to show, something I never saw again, and something I have thought about ever since. The scene was a room about the size of an average bedroom, but with padded walls(!). A man was standing about six feet from the back wall, from the camera's perspective. Another person was facing him, maybe ten feet away. This second person moved his right hand up in front of him like he was holding a cup (in terms of his hand position), but his four fingers were protruding through a device. I could also say that the device was wrapped around his fingers. This device was about an inch across and maybe half an inch thick, shaped like a three-dimensional letter 'O', with its hole big enough for his fingers to fit through.

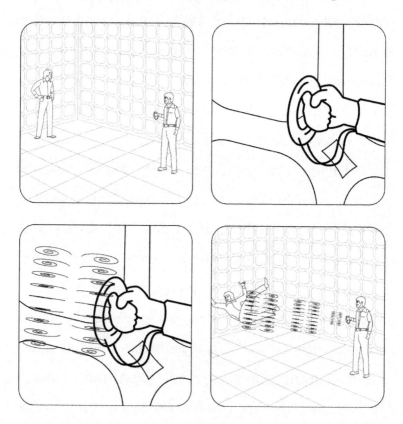

It was turned on somehow, and the following happened rapidly. Right in front of the device the air became distorted, similar to air distorted by heat, in the shape of two vertical back-to-back spinning coils.

This spinning form progressed forward towards the first man, enlarged in size, and when it reached him, knocked him off his feet to the padded wall behind him, rather forcefully. I have to stress that it didn't appear to be the air spinning or moving, but some distortion, like two vertical coiled waves moving through the space. Needless to say, I was taken aback and woke my wife, but it was only shown once. I looked at the news and scanned the papers for several days, but never saw the story again, or anything like it. Over time I came to believe that the station aired a story that slipped by a censor or was leaked to them, but which 'shouldn't' have aired. In the end though, only a small number of insomniacs saw it, and it might have registered with a miniscule number of those, maybe only one.

I relate this story to bring up the fundamental premise of this book, that there exists a universal substrate of all things, called by various names over time, including the aether. I will refer to it as ether. It is the thing of which the electron, proton and neutron, and all other 'particles' are made. It is present in the 'vacuum' of interplanetary space and it permeates every thing everywhere in the universe. I believe that this TV news story showed a device that affected (moved) the ether by some electromagnetic means, and by moving the ether, caused the man to be thrust backwards against the padded wall.

I realize that by most accounts the concept of the luminiferous ether is dated and not accepted in the scientific community. Since Einstein showed how a packet of energy could propagate through empty space on its own, the concept of an ether has not been needed to explain the standard theories in science. But I choose this view because I believe that it best represents reality, and because it can unite concepts such as gravitation, electricity, magnetism and others. When I start to feel old-fashioned, I remember Nikola Tesla resisted more modern views and used his ether-centered view to almost singlehandedly create the alternating current polyphase motor/generator system that has powered the world's grids for over a hundred years. Almost as a side note, he is also responsible for radio, remote control, fluorescent bulbs, stranded wire and so much more. Indeed, if he had his way, we would all be drawing free electric power from simple boxes attached to the ground, or from antennas in the air.

I posit that the ether is real and part of nature, although it is almost entirely undetectable directly. It is the 'other', whereas everything that we see - the planets, solar systems and galaxies - are considered to be 'normal' matter. What we do experience are the patterns of movement of the ether - gravity, electromagnetics, atomic bonding, and many other phenomena.

This work is a humble attempt to integrate the ideas of people like G. Patrick Flanagan, Wilhelm Reich, Nikola Tesla and many others, for the purpose of developing a coherent and workable understanding of the universe, and applying this understanding to solve today's problems. It is not written as a scientific text and has few formal citations. My method is to present a line of thought, make an argument, and illustrate it with what I believe to be true, or at least most probable, facts. The internet makes it quick and easy for readers to verify for themselves the validity of anything presented here. If something in this work interests you, I would greatly encourage you to explore varied sources for more information.

This work is written in a format called 'ventilated prose', 'borrowed' from Pat Flanagan. This format, or style, gives the reader a moment to think about the complete thought presented before moving on to the next idea. I hope it works for you.

CHAPTER ONE:
THE ETHER

Properties of the Ether

The ether is the cause, and its movements and patterns of movements that we perceive are its effects. My attempt here is to describe the properties of the ether, called by many names throughout the centuries by its many discoverers, and to provide a logical progression of thought that will, hopefully, lead the reader to an alternative way of understanding nature. In Chinese philosophy, Chi (Qi), or life-energy, flows throughout the human body in pathways just below the surface of the skin, and is accessed and adjusted via low-resistance acupuncture points. Called Ki in Japan, this substance also completely fills the universe and is what connects all things everywhere. In Indian studies, Prana flows into the body with the intake of breath, and Kundalini is taken in from contact with the earth, forming a life-long pulsation of energy progressing up and then down the spinal cord. Polynesians call this energy Mana, and Karl von Reichenbach studied what he called the Odic Force.

And as we see here, Wilhelm Reich extensively studied what he called Orgone energy.

Dr. Wilhelm Reich was a colleague of Freud and Jung, but branched off in the direction of natural science. He discovered what he called Orgone in his laboratory, which he said was a mass-free energy with no inertia or weight, found everywhere in the universe, but in variable concentrations. Huge quantities of it flow out from the Sun and pass through all matter, but at different speeds. He found that Orgone displayed negative entropy – flowing from lesser to greater concentrations of itself, and is absorbed by water, refracted through a prism and reflected by polished surfaces. According to Reich, the thin blue envelope surrounding the Earth is Orgone energy, constantly moving west to east faster than the speed of the Earth and its atmosphere, expanding and contracting along the path of its motion. Somewhat controversially, he once compared the properties of this energy with certain qualities of a Supreme Being.

The energy is everywhere: it fills space as far and as close as we can observe, even in a high vacuum, and is eternal, having no beginning and no end. In addition, it's the source of all life and consciousness.

Density Levels

The ether (orgone) is present everywhere, but at different density levels. The space around a planet has much more energy in it than an area say, outside the solar system, but nowhere can be found 'empty space'. Not only are there higher density levels in the area around a planet, but in localized areas there can be different density levels in the same area. This concept of energy systems sharing the same space even provides us with the framework for a multidimensional universe. The ether in an area can be at many different levels of density, each vibrating and moving in their own sets of patterns. If the space in your home were also filled with a much finer density level of ether that would correspondingly vibrate at a much higher rate, it is possible that this energy system could function at this higher frequency without normally affecting other systems in another dimension in the same space.

Pulsation

The ether exhibits pulsation in the direction of its motion. In some energy systems it expands and contracts in rhythm, with its frequency based mostly on the size and density of the system. A stream of energy is a series of high and low density levels spaced at regular intervals, somewhat analogous to a sound wave, which has regularly spaced waves of compressed air:

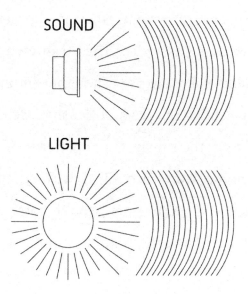

SOUND

LIGHT

Vibration

The ether carries electromagnetic vibrations of the universe from one place to another. It is out of fashion now, but it was once referred to as the luminiferous ether, meaning it carries or bears light. If you think of light as a wave, then the ether is what waves. Sunlight is an example. Reich viewed sunlight in terms of its effects on Earth, in an overall pattern of energy flow in the solar system. According to Reich, the energy of the solar radiation travels out through space, where the density level is very low, so little of the energy is lost. When the vibrations of the solar radiation reach the Earth, where the energy density is greater, more of the solar energy is absorbed – in the atmosphere, the seas, the land, and almost all living things.

With an ether theory, electromagnetic radiation is viewed as packets of waves of energy (quanta), traveling in space at the speed of light, with a certain frequency, amplitude and waveshape. Because the continuum of ether is the medium for this traveling wave of energy, the amount of locally dissipated energy in a wave is dependent on the etheric density of the area. And because the density of the continuum of ether is very low in 'empty' space, very little of the radiation is expended in this space. Additionally, resonance in frequency between the traveling wave and a local energy system can cause the wave to either gain or lose energy.

Spinning Wave

Reich's model of a traveling wave of energy is a "spinning wave" (Kreiselwelle) (Reich, 1973), which spins in a perpendicular direction relative to its line of progression. Without knowing it, mathematicians and physicists have graphically represented this 3-dimensional spinning wave 2-dimensionally as a sine wave:

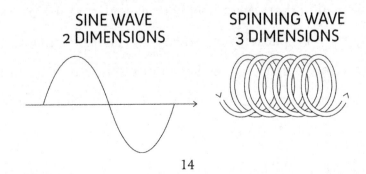

SINE WAVE 2 DIMENSIONS   SPINNING WAVE 3 DIMENSIONS

Negative Entropy

A large energy system tends to break down to smaller and smaller systems. This is entropy, on display in many material processes. A rock is thrust in a river, where it is gradually worn down to its individual mineral components, which are then washed down to dissolve in the ocean. The ether displays negative entropy – it is always attracted to larger concentrations of itself. A cloud, which is both a local concentration of water vapor and of ether, is an example of this quality of negative entropy. Consider a quiet enclosed valley in the early morning. Tiny drops of water condense on plants because of the temperature difference between the cool air and the plants connected to the warm Earth. These drops of water evaporate and create a bank of mist, held down in this valley between the opposing hills by its heaviness relative to the surrounding air. The Sun rises and the moist air is heated by absorbing the energy from this star, which has traveled about 93 million miles in 8 minutes. The molecules in the bank of moist air now get hotter, and more heat means more movement. This heated volume pushes out and expands, as all heated things do. By expanding, the bank of moist air becomes less dense (lighter) than the surrounding air, so it begins to rise. A cloud is formed.

And here is the critical part: If entropy were the only force working here, then this cloud would forever expand until it was more or less evenly distributed in the surrounding air. Instead, the cloudlet is heated and expands, but it constantly attracts more moist air into itself. The cloud expands and condenses until it reaches equilibrium with the surrounding air, finally floating at a more or less consistent altitude. It is now made up of more and more flow patterns of swirling eddies, which functionally sweep in and condense the moisture out of the air at the peripheries of the cloud, providing the cloud with more moisture to be heated and to expand. The attraction to itself keeps the cloud together, as does its own set of flows, circling and curling in on itself.

Buckminster Fuller, the inventor of the geodesic dome, expanded on this idea by proposing that a geodesic sphere of at least a half-mile in diameter made of struts and hubs could, by reflecting sunlight multiple times back inside the sphere, heat the interior air and force enough of this heated air out of the structure to offset the weight of the structure, causing it to exhibit buoyancy in the air – to float. As the size increases significantly from here, any weight added to the structure in terms of floors, furniture, etc., becomes negligible compared to the weight of the additional expelled air. Fuller envisioned giant orbs containing everything needed to live, with populations of people floating with the wind from mountaintop to mountaintop, only tethering when necessary. Even now a project like this, huge as it seems, is possible, given the vast resources of a large corporate entity, or of some ambitious government.

Cloudbuster

Wilhelm Reich invented a device which he called the 'cloudbuster', which could pull a stream of energy into itself. He found that the orgone has a strong attraction for water, especially running water. He ran an ordinary length of wire into a running stream and attached the other end to a block of wood, to which was attached an array of metal pipes pointing out from one face of the wood. The modern theory of this phenomenon is that the running water draws ether from the wire, which causes a draw of ether from the block of wood. Ether from the atmosphere is pulled into the wood block through the metal pipes. Long pipes laid out in a hexagonal array seemed to work best, causing a straight stream of ether to be sucked into the device. When the cloudbuster was pointed right at a cloud, a stream of ether from behind the cloud would blow the cloud apart on its way to the device, thus its name. However, if it were pointed just past the edge of a cloud, the cloud would expand, because some of the induced stream would be attracted to the cloud and get caught up in its flow.

Reports emerged in the 70's of Russian weather engineering efforts using some of Reich's concepts, but those reports, sketchy as they were, seemed to conclude that the law of unintended consequences prevailed. When weather patterns are modified by artificially altering the flow of atmospheric ether in one region, an adjacent region might experience an unintended outcome, either negative or positive. One would like to think that lessons were learned and the efforts ceased.

Duality

Another quality always present in the ether is its duality, expressed most clearly in the famous Yin-Yang symbol:

The Yin (shady side), expressed as black, encompasses the female, inward motion, while Yang (sunny side) is white, male and outward in motion.

The two complement and complete each other into a unified whole. Notice that each symbol, Yin and Yang, expresses its own complementary parts, white and black, to complete itself. Nothing in nature is pure Yin or Yang; each contains the seed of the other. Additional expressions of this Yin-Yang duality:

Yin	Yang
Shady North Side of Hill	Sunny South Side of Hill
In	Out
Female	Male
Cold	Warm
Inhibition	Excitement
Dark	Light
Intuition	Logic
Earth	Heaven

There are times when the Yin dominates, and times when the Yang manifests itself more strongly. This basic natural concept is manifested in the seasons, in plant and animal cycles, and in all aspects of life. The living body breathes in (Yin), then breathes out

(Yang), and the breath is the energy (Chi). The yearly cycle of the Earth's orbit around the Sun, relative to each hemisphere, progresses from the maximum Yang of the summer (fire, red), to the autumn (metal, white), then to the maximum Yin of the winter (water, black), and finally to the spring (wood, green).

This duality expresses itself often in nature as forces in opposition to each other. If it's said that nature abhors a vacuum, which it does, it must also be said that it can't abide an imbalance. When the area at the bottom of a thundercloud becomes positively-charged, a large buildup of negative charge (electrons) forms in the ground under the cloud. If the charges are great enough, and if the distance between them is small enough, leakage occurs in the form of lightning, where electrons, in this example, travel from the earth (electrical ground) to the cloud. Another example of this principle of complementary formations, but with a completely different pattern of movement of the ether, is when the north pole of a ferrous magnet is in contact with an iron nail. If kept there long enough, the part of the nail in contact with the north pole <u>becomes</u> the south pole of its own little magnet. The north pole here <u>creates</u> a south pole. The principle is that if there is a great enough concentration of an energy, then its opposing, or complementary energy is formed nearby.

The negatively charged earth (ground) is complemented by the positively charged ionosphere, the area of our atmosphere from about 60 km to 1000 km, and the entire terrestrial system functions acts as a giant electrical capacitor, essentially two charged plates of opposite polarity separated by an insulating material. The insulating or dielectric material between these 'plates' is air, thin as it is here, and includes the blue layer of ether at about 330 km. A capacitor holds an energetic charge between its plates, and the amount of charge is a function of the voltage between the capacitor plates, and its geometry. The charge held between the ionosphere and the Earth is none other than the ether, an energy held in place as long as the positively-charged ionosphere complements the negatively-charged Earth. This duality, or system of complements, is seen time and again in so many aspects of our experience – male/female, light/dark, etc. The following summary of one researcher's experiments sheds some light on this duality.

Reichenbach's Experiments

Baron Karl von Reichenbach conducted his research on the effects of the ether, which he called Odic force, in the mid-19th century. Although by our standards today he might be criticized for using unscientific methods and employing 'sensitives', or people especially sensitive to these effects, to his credit he repeated his experiments over and over, with his sensitives recording similar perceptions each time. As a group, these sensitives had common traits, such as: avoiding the color yellow and being drawn to things blue, disliking mirrors, sleeping only on their right side with their heads pointing north, avoiding crowds, and generally eating plain, cold and simple foods. The main experimental results pertinent to this work are presented here (Reichenbach, 1852).

Large crystals were placed on the corner of a table, and the sensitives all felt a coolness emanating from the pointed end, and warmth from the blunt end. In a darkened room they saw a bluish light coming from the pointed end, and reddish-yellow light coming from the blunt end. When a bar magnet was placed on the corner of a table, the coolness came from the north pole, and warmth from the south pole. The color blue was associated with the coolness from the north, and red or reddish-orange with the south. Once again, the pointed end of a crystal and the north pole of a magnet displayed coolness and the color blue; the blunt end of a crystal and the south pole displayed warmth and a reddish color. I interpret these observations from an etheric point of view. The north magnetic poles and pointed ends of crystals exhibit an outward flow of the ether (yang), while the south magnetic poles and blunt ends of crystals have inward etheric flows.

The polarities of his Odic force were shown further when luminous strands were seen emanating from fingertips in a darkened room. From the right hand came a cool, blue light; from the left came a warm, orange one. When two sensitives grasped each other's right hand (like pairing), they perceived an unpleasant warmth. If they grasped each other's left hand with their right, and right with the other's left (unlike pairing), then the effect was a pleasant coolness. One other point of special interest in Reichenbach's experiments is that when direct current electricity was conducted through a wire in a darkened room, it was seen to be accompanied by a "corkscrew kind of light which travelled rapidly around it".

Movement as a Force

Movement of the ether comprises a force; in fact, a force is defined as a movement of this ether. What else can a force be? Wilhelm Reich thought of his orgone energy as being in constant motion, but mass-free, with no inertia or weight. Relative to human awareness, it is extremely finely distributed, but it must have some mass, even if it is only an effect of its motion, for it is indeed never still. This force acts on material objects according to Newton's formula $F = Ma$, which states that the force on an object is equal to its mass times its acceleration, or change in velocity. In functional thinking, a stream of ether (force) impinges on an object in some way, and causes that object to move, or in some way change its inertial system (think of the man in the padded room). The equation merely gives the quantitative ratios of the concept, while an understanding of the functioning of the system leads to a better conceptualization in one's mind.

Spirals

The ether tends to move in a curved path rather than a straight one; in fact, its major pattern of movement is the spiraling motion, examples of which abound in nature. The rolling of an ocean wave is one: an isolated particle floating on the water undergoes a motion of rolling in a circle while progressing in one general direction – the spiraling motion. The DNA helix, crystals, sea shells, plants, ocean currents, tornadoes, hurricanes, solar systems, spiral galaxies and even black holes are examples of the twisting energy that guides the growth of all living things – animal, vegetable or mineral.

An interesting note here is that phosphenes, which are the points and shapes of light perceived when the eyes are shut, have definite shapes more or less the same from person to person, and the spiral is one of the more basic and universal forms. Does the eye, in total darkness, directly perceive and recognize the spiraling of the ever-present ether? Nikola Tesla, in my mind the creative genius of the age, found in his work with resonant transformers that the most elegant and efficient form of a transformer winding was the flat spiral, which concentrated the electric portion of the electromagnetic (etheric) flow to a greater degree than any other form.

Orgone Accumulator

The ether's ability to concentrate itself in a given space is another of its important qualities, and the clearest illustration of this that comes to mind is Wilhelm Reich's Orgone Accumulator. This was a unique and original inspiration of this creative natural scientist. Its operational functioning is produced by a layering effect. A layer of metal is alternated with a layer of organic material, such as wood. In an accumulator, the inside layer is metal, and the outside always organic. (It can be argued that the inside layer is air, which is non-conducting. This is where the real accumulation takes place.) Reich's theory is that organic material attracts ether and absorbs a certain quantity of it, while metal also attracts it, but immediately repels it. In any layered wall, the net flow of ether is towards the side whose last layer is metal, as shown in this drawing:

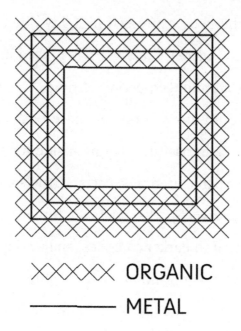

$\times\times\times\times$ ORGANIC

———— METAL

As stated, in an accumulator box the inside area is entirely surrounded by metal, and the outside surrounded by wood (organic). The net energy flow on all sides is towards the inside, and this is where the concentration of ether occurs. The concentration is limited by the maximum amount of energy which the space and the material can hold, but it can be quite large and readily noticeable to human awareness, as recorded in Reich's experiments. Basically, the effects on humans sitting in a comfortably-sized accumulator box are a charging effect on the body expressed in many different ways, and a general increase in vigor when used in the right doses over a prolonged period of time. The only harm a person could do with one of these boxes would be to overcharge himself by staying inside for too long a time, past when common sense and one's own senses tells one to leave. Sparks, mists and other light effects were reported in darkened boxes, suggesting charging of the energy to the point of local illumination. The colors seen were mostly bluish-gray or bluish-green, seemingly validating Reich's association of the energy with the color blue.

The connection between the ether and the color blue was personally reinforced some time ago when I witnessed the nighttime explosion of a large neighborhood electrical transformer. It was as if a box filled with intense blue light suddenly burst open, momentarily bathing a hillside in blue. This makes sense if the ether affects and is affected by electricity, as we shall see in the following chapter. An electrical transformer is a device that concentrates and transfers large amounts of electromagnetic energy in a small volume of space, and when one explodes, the associated blue energy of the concentrated ether escapes and quickly dissipates into the surrounding area.

This effect was again confirmed in December 2018 when a large transformer in the New York City borough of Queens exploded at night and produced a momentary burst of intense blue light, witnessed by many and recorded by the news media. It occurs to me that an area for future research would be the investigation of the precise blue color (frequency) of the light recorded in these events as a clue to the true nature of the ether.

Dark Matter

This chapter could not be complete without a discussion of the two most important discoveries of recent times, dark matter and dark energy, although we really just 'discovered' how much we don't understand about these fundamental components of the universe. Dark matter, which comprises about 27% of the entire mass-energy of the universe, has been theorized for centuries as the missing piece of the universe. Observations and equations seem to demand this missing piece be accounted for.

The existence of dark matter can be inferred by several different methods, but the following evidence is strikingly powerful. When galaxies are observed, they exhibit an odd phenomenon. To understand it, picture water in a sink swirling down the drain. The water nearest the drain spins faster than the water further away. In the same way in our solar system, the innermost planets revolve faster around the Sun than the outer planets. (Technical note: both the Sun and the planets actually revolve around the common center of gravity of the solar system.) Tornados, tropical storms, and even black holes seem to exhibit the same behavior, but not galaxies or galactic clusters! As a rule, galaxies function more like an LP record (CD disk for the youngsters) than a swirling vortex of liquid or gas – the stars near the center of a galaxy have about the same angular velocity as stars in the outer fringes of the galaxy.

The direct result of these galactic observations was the further advancement of the dark matter theory. Indeed, if the measurements and equations are close to being correct, to account for these galactic observations there needs to be about 5 times the amount of some unseen mass than there is normal matter, which is basically everything made of protons, neutrons, electrons and neutrinos. So a galaxy has to have a lot more mass than what is observed, a dark matter, which only interacts gravitationally with normal matter. By the laws of physics as we know them today, this is the only real answer to these curious galactic observations.

Current maps of the universe as far as we can see show dark matter not being evenly distributed, but clumping together in incredibly long filaments, with gravity as the driving force at this macro-level. And 'normal' matter (galactic clusters, galaxies, solar systems etc.) tends to congregate along these filaments. It is not known if this normal matter is gravitationally attracted to volumes of higher dark matter concentration, or if dark matter naturally concentrates around normal matter. Presently, there does not seem to be any direct connection between dark matter and the ether, but dark energy might be a better candidate.

Dark Energy

Like dark matter, dark energy is said to be dark because of all that is not known about it. The best measurements to date show dark energy to comprise about 68% of the total mass-energy of the universe, even though it is said to exceedingly thin, yet almost uniform in density ($\sim$7x10^{-30} g/cm^3 by Flanagan's estimate). Observations show that galaxies are not just moving away from each other, but amazingly, the speed of this expansion is steadily increasing, The expansion is accelerating. To explain the observed expansion and acceleration of the universe, a constant negative pressure, a gravitational repulsion, must exist.

Dark energy fills all of space, even vacuums, but can have different local density levels. It vibrates and pulsates, is said to affect normal matter only gravitationally, and appears to be the cause of the acceleration of the expansion of the universe. The common characteristics between what I've described as the ether and dark energy are obvious. Could dark energy be the substrate material of the universe, the thing that flows in patterns called electricity and magnetism, and atomic and subatomic particles? I don't know, but I believe more and more that there is some connection between the two concepts.

CHAPTER TWO: FORMS

Electron, Proton and Neutron
Gravity
Static Electricity
Electromagnetism
Electronic Components
Magnetic Coils
Bifilar Coils
Floor Plan at Chartres
Nazca Lines
Ley Lines
Human Ether Flows
Acupuncture
Auras
Rosicrucian Order
Church Architecture
Feng Shui
Dowsing

Electron, Proton and Neutron

The major characteristics of this universal ether were named and briefly described in the preceding chapter, while this chapter shows the many forms, movements, and patterns of movement of this ever-present energy. The following section on elementary particles describes specific patterns of its movement. Every event is an energy event, and every 'particle' is a self-repeating pattern of etheric movement. To explain the basic forces of the universe and atomic particles, I turn to the work of Dr. G. Patrick Flanagan, whose chapter on his theory of energies in Pyramid Power unifies many concepts behind natural events (Flanagan, 1975). The concepts presented here are do not represent the current views on subatomic structures, but do introduce the reader to the torus, or donut shape, which seems to be a strong candidate for a possible self-repeating pattern of movement of the ether. The electron is pictured by

Dr. Flanagan as a point-to-point double vortex of ether:

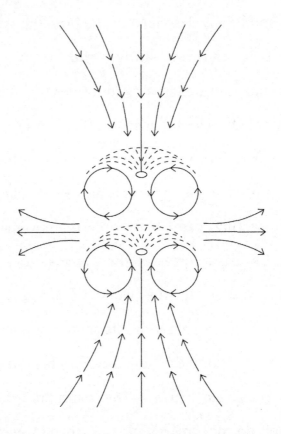

Ether comes into the electron through the opposing poles and goes out through the equatorial section. A kind of back-to-back rotating ring pattern forms, where the surfaces of the rings roll with each other. The electron is an entrained flow of ether thus described, but it is not an isolated system. Ether emitted from the equator does not all curl around to go back into one of the poles again. Some escapes, but it is balanced by ether from its surroundings coming into one of the poles. The electron, in essence, is two directly opposing balanced streams of energy colliding head-on and squirting out the only way left - the equatorial region. This causes back-to-back toroid shaped structures, defining the different areas of the electron.

According to Flanagan, a proton can be thought of as a reverse electron; energy enters at the equator and shoots out the poles. Because it enters through the large equatorial region, more energy can be entrained, accounting for the greater mass of the proton. The essential flow of a proton is two streams of energy moving away from a point in exact opposite directions. Energy comes out of the poles and most of it goes back into the equator. The rolling rings of energy spin in the opposite direction:

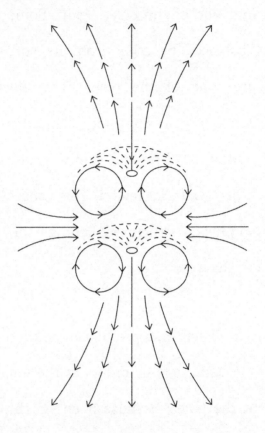

The polarities of protons and electrons can be identified by their polar flows. The positive polarity of the proton is the flow out of its poles. The attraction of a proton to an electron in a hydrogen atom can, with this theory, be illustrated as a functional flow of energy:

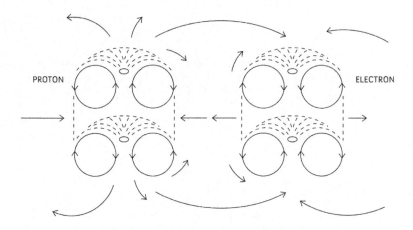

I have not described this as electric or magnetic coupling, but rather simply a flow of energy between two 'particles'. With this flow, the electron is free both to revolve around the proton and to spin, which we know occurs.

The neutron is comprised of three rings of energy rotating back-to-back-to-back, which, if we assume the surfaces of adjacent rings to always be in rolling contact, necessitates a neutral polar flow. The flow into one of its poles is balanced by the flow out of its

opposite pole:

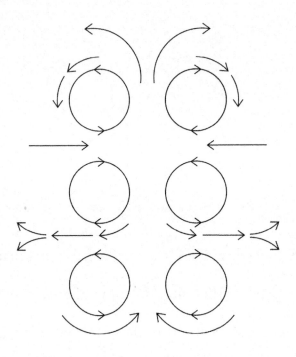

This was Dr. Flanagan's vision of the elementary

particles in 1975. The current theory is that there are

quarks and leptons, in different varieties, among

other basic subatomic particles. The forms or shapes

of quarks are unknown. Two 'up' quarks and one

'down' quark combine to form a proton. Two 'down'

quarks and one 'up' quark form a neutron, and both

the proton and neutron are in a class of particles

called hadrons. The electron is an elementary particle, one of the leptons, and again has an unknown structure. The incredibly small scale of these subatomic 'particles' renders direct observation impossible today, highlighting the difficulty in a real understanding of this subject. The basic torus, or donut shape, with energy spinning in and out, remains a good candidate building block for these ultra-small particles.

If the three elementary particles are no more than specific patterns of movement of the ether, then what is solid matter? What we perceive through our senses as a hard wooden table is simply a set of fast-moving, very small whirlpools of energy, bound by their common flows to other 'particles', giving the appearance of solidity. This has been called the 'Grand Illusion', among other things, and reminds us that a real understanding of something in nature requires more than our human senses. The true amount of energy involved in these flows is shown most clearly and most violently in atomic fission. Blowing apart the bonds of energy in what we call a small piece of matter releases the energy that we associate with nuclear weapons. Radioactive elements work best because they are already overcharged with energy.

Gravity

In everyday use, the only thing about gravity that we really know is that we feel a pull down against the surface of the Earth. As children we learn naturally that if we fall off of a tall enough object, our speed when we hit the ground can be great enough to hurt us. Cavemen knew that if they could dislodge a large rock from a steep cliff, then what they understood to be gravity would cause that rock to roll down to a lower point. They used gravity to the extent that they understood it.

Sir Isaac Newton gave us the next jump in our understanding of gravity. From Newton we learned that every mass in the universe attracts every other mass, and that this attraction is proportional to the product of the size of the masses, and inversely proportional to the square of the distance between them:

$$F = G \times (M_1 M_2 / d^2)$$

The force holding us to the Earth is equal to the gravitational constant times the product of the mass of the Earth and our own mass divided by the square of the distance between them (really between us and the center of the Earth). This knowledge gave us quantitative values of the force of gravity precise enough to suit the needs of a growing technology, including propelling and guiding spacecraft to the Moon, and beyond. However, the disadvantages of almost an exclusive use of a purely mathematical understanding are twofold. Because it is purely mathematical, it heavily stresses the analytical side of understanding rather than a unified or holistic one. It also introduces us to the concept that was held by science for hundreds of years, that the space between planets and their sun, and between solar systems, is for the most part empty space, devoid of almost all matter and energy. The evidence today, to understate it, strongly favors a universe filled, without any observable interruption, with a basic energy of many different density levels, vibrating at every imaginable frequency. Strange particles, quantumly joined to other particles across the universe, pop in and out of existence, and 'ordinary' matter only accounts for about 5% of the universe.

Albert Einstein's view of 'curved space' was the next major development in our understanding of gravity. Where Newton saw gravitational attraction as a pull from center of a large mass, Einstein saw every mass bending the fabric of space toward its center. An object traveling near the Earth is bent toward it because space itself is bent around the Earth. Empty interplanetary space is assumed, but that space's composition has bends or warps built into it around massive bodies, and these warps in space are carried along with the body in any orbital movement. The concept of a gravitational 'field' is built into this outlook; a volume of space starting at the center of the Earth and radiating out is affected by the gravitational field of the planet, to a degree based on the mass of the planet and the inverse of the square of the distance. In Einstein's view, gravity is a property of space itself, and the more mass that there is in this space, the more gravity there is.

In an etheric universe, gravity is the pressure of the ether. Rather than the concepts of mass attraction or curved space, Dr. Patrick Flanagan stated that it is the pressure of the surrounding and interpenetrating ether that presses us against the crust of the Earth. We are held down not by the pull from the Earth, but by an overall net pushing down of the ether (its weight) on top of us. As we gain altitude, our own weight naturally decreases, because there is less concentration of ether above us, and greater concentration below us.

Again according to Flanagan, the Sun and the Earth do not attract each other. Rather, streams or jets of ether naturally flow out in all directions from both bodies, but the streams in the area between the bodies collide with each other and shoot off in a direction roughly perpendicular to their original course. This creates a lower pressure in the area between the two bodies, which has to be complemented by a higher pressure on the opposite sides of the bodies:

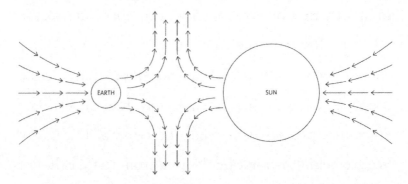

With this pressure system, the two bodies fall in toward each other, and would collide if not for the centrifugal force of the orbital movement of the Earth. This ether theory helps us to explain gravity without the concept of 'action at a distance', which has hampered our understanding of the subject for centuries, and had baffled Newton so much that he basically deferred it for future consideration. It is a qualitatively different way of viewing gravity, yet it violates none of the end results of either Newton's or Einstein's ideas.

Gravitational waves have been theorized for decades, along with the question of their speed as they propagate through space, thought to be the speed of light. In 2017 the light from an explosion of two neutron stars colliding 130 million light years away was detected, along with its faint gravitational wave registering at the same time. The importance of this detection event is that it provides extremely strong evidence, if not proof, that gravity waves do indeed travel at the speed of light, as predicted. Of course as with light, the question of whether it is a wave or particle (graviton), continues. A graviton can travel through an empty space, where a gravity wave needs a medium to travel through, something to 'wave'.

Let me state that no one today has an absolute understanding of the nature of gravity. No one has experienced fully the thing we call gravity; we can only see or feel its effects. We see a chair resting on the floor and assume that there is some force below it pulling it down. Likewise, we feel our weight when we jump in the air and feel a strong tug down. It is just possible gravity is not a pulling-down force from the Earth, but the pushing down of the ether from above. The effect is the same, but the concepts and the implications are vastly different, and, as we shall see in the final chapter, we may never build a true anti-gravity device if our ideas about gravity remain the same.

Static Electricity

While gravity is the overall pressure system around a large object, electricity and magnetism are much more local patterns of movement of the ether, and act pretty much independently of the general concentration or local density level. Static electricity is really a local concentration of electrons, with a negative charge, matched by a nearby build-up of positive polarity, or lack of electrons. It is static. If no other conditions change, then the two charges that are separated by a non-conducting material will remain the same, except for some small leakage of the charges over time. The potential voltage is the pressure of one concentration against the other, and is equal to the amount of energy that would flow if a conductive path were provided.

A local build-up in a thundercloud becomes too great, and the space between the mostly positively charged cloud and the mostly negatively charged Earth can no longer hold back the resulting lightning bolt, in which energy is transferred both ways in order to restore balance of polarities. The rarely seen ball lightning, or St. Elmo's fire, is a local build-up of positive energy so concentrated that it naturally forms a sphere, and holds itself together in its mini-gravitational field. The pressure system around the sphere acts for the most part independently of the general gravitational pressure; it floats, and seeks to discharge itself at a point of high negative polarity, with enough energy transferred to kill a human.

Electromagnetism

In a storage battery, we know that because of the chemical action between the liquid acid and plates, a charge of electrons accumulates on what we call the negative terminal. The positive terminal has a lack of electrons. These electrons do not move from the negative terminal of the battery because the resistance of the surrounding air is extremely high. When a path of small resistance (wire) is connected to both terminals, electrons flow from the negative terminal around to the positive, and flow around and around until the acid in the battery is neutralized, due to corrosion and the imperfections of the materials used. We know that the electron flow pressure (E) is equal to the amount of flowing electrons, current, (I) times the resistance (R) of the medium of the flow. This is Ohm's Law: $E = I R$.

In a wire with electrons flowing steadily in one direction (direct current, or DC), there is a steady flow of ether around the wire:

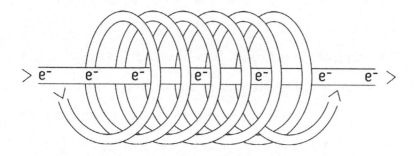

If we take a long wire with electrons flowing through it, and coil it many times around in a small area of space, an electromagnet is constructed, and this provides an understanding of the flow of ether in a normal ferrite magnet. An electromagnet and a ferrite magnet both have a north and south pole, and one pole of either will attract the opposite pole of the other, so the flow patterns of the ether around each must be, at the least, extremely similar - they are both magnets. In an electromagnet, the electron flow and the magnetic (etheric) flow are illustrated by the left-hand rule of electromagnetic theory. If you curl the fingers of your left hand and stick your thumb out, your fingers represent the coils of wire and the direction of the flow of <u>electrons</u> in the electromagnet. Your thumb represents the etheric (magnetic) flow of ether out of the <u>north</u> pole of the electromagnet. Note that this is electron flow, not 'conventional current flow', which is really the progression of positively charged 'holes' in the wire left by the movement of electrons. A simplified way of understanding almost any electrical device is that it is something that uses a voltage (pressure or force) to pull electrons from ground, and perform a useful function on some resistive load (heater, motor, etc.). A vivid memory from an electronics class session was a discussion of current flow in transistor circuits, which was for me a confusing subject. Our teacher ended the confusion by telling us to always follow the electron flow, because that's the only real thing moving in an electronic circuit.

In a real electromagnet, the wire is coiled around a ferric rod, which concentrates the etheric flow. This overall etheric flow pattern of an electromagnet is the same as a ferric magnet, where ether flows out of the north pole and back around into the south pole:

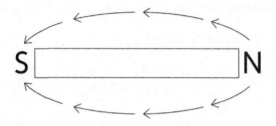

The answer to the crucial question of why opposite poles attract and like poles repel cannot be answered by any other concept but this. When the north pole of a bar magnet faces the south pole of another magnet, the flow of ether is in the same direction through both of the magnets:

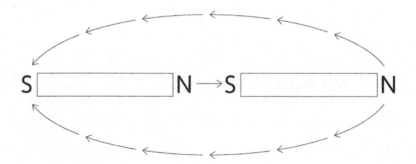

The flows agree, and the north and south ends are drawn toward each other. Energy (ether) is transferred from one magnet to the other, traveling through the magnets in the same direction. Now when like poles face each other, the flows through each magnet are in opposite directions, so the magnets repel each other:

S[   ]N —→ ←— N[   ]S

It is superfluous to talk about a magnetic flow or an electric flow separately. Where one exists, the other must also. In an electromagnet, the ether spins around the individual wires in the coil, causing ether to flow into the South pole and out of the North pole of the electromagnet. I believe the system described here is a close approximation of the true pattern of flow known as electromagnetism.

Electronic Components

Let us look at the basic electronic components from an etheric point of view. In an electronic circuit with a battery providing DC electron flow, some resistance device impedes this flow, whether it is a resistor made from carbon or ceramic, or something else doing work and creating heat, noise, etc. The key is the heat, which is really motion, for where there is heat there is a local concentration of ether. A capacitor, or condenser, at its essence, has two metal plates with some insulating material separating them. One plate becomes negative in polarity, meaning electrons amass here, and the opposite plate loses electrons, becoming positively charged. Its capacitance is the amount of charge held in the insulating, or dielectric, material between the charged plates. In some circuits these opposing charges build up and remain on the plates, while other circuits have plates that shift back and forth in charge. In capacitors, while electrons gather on the negative plate, the ether is attracted to the positive plate.

Magnetic Coils

A crucially important phenomenon in the study of electronics is that a wire with electrons flowing in one direction induces electron flow in a nearby parallel wire in the opposite direction. In an electromagnet, there are many parallel wires typically wound around some type of iron core to concentrate the lines of magnetic flux and to increase its strength. Each of these windings of the wire has electrons flowing in the same direction, so each winding induces the electrons in its neighboring windings to flow in the opposite direction, creating a 'back' voltage that works against the applied voltage in the coil. In simple terms, the battery or other power source has to work hard to power an electromagnet.

Bifilar Coils

There is another type of electrical coil, called bifilar, which works differently. Bifilar means two filaments, or wires, and these coils are used in specialized electronic equipment. In a typical electromagnet, a single wire is coiled many times in the same direction around a cylindrical shape, and a direct current (DC) is applied. Magnetic force is created, meaning ether is made to flow out of the north pole, and back around to enter the 'magnet' through the south pole. A bifilar coil can be made with stereo wire, or two insulated wires joined together. The two wires are stripped on one end and soldered together, then the double wire is wrapped around a cylinder as in the usual electromagnet. The difference is that each wire is next to other wires with current flowing in the opposite direction, causing the magnetic fields to cancel one another. And since each wire is adjacent to wires with electrons flowing in the opposite direction, the electrical field is enhanced, with no back voltage. Again in simple terms, the power source for a bifilar coil does not have to work hard to produce its effect. Tesla discovered and patented a flat spiral bifilar coil that he found to be more efficient than any other coil.

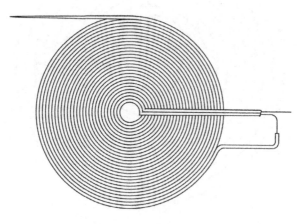

Floor Plan at Chartres

I bring this up to show a possible connection to something seemingly unrelated. I have been fascinated with the floor plan for the Cathedral of Chartres in France since I first saw it. In a distinctly modern practice, pilgrims today slowly walk this labyrinth in prayer as a devotional. What I find most interesting is that if we imagine this labyrinth filled with people walking through it from the outside into the center, we note that everyone is always surrounded by people walking in the opposite direction:

Nazca Lines

Chartres is far from the only example of this phenomenon. The Nazca Lines in Peru contain several figures that can be only described as flat, spiral bifilar coils. In addition, several animal figures are composed of back and forth weaving lines. If people walked any of these lines in a procession, as we believe they did, everyone would again be surrounded by people walking in the opposite direction:

It has been pointed out by many authors that the Great Pyramid at Giza contains in its structural dimensions references to mathematical constants such as Pi and Phi (symbol for the Golden Proportion). Were the builders simply showing off their mathematical knowledge, or were they trying to pass this knowledge on to future generations? In the same sense, were the designers of Chartres and the Nazca figures trying to point to some technological principle, possibly electric, knowing that like the pyramids, their work would last for hundreds or even thousands of years? It's a stretch, but not so much in my view.

There is also a figure on the Nazca plains that I find particularly fascinating. It's called the zig-zag pattern, and I can't help seeing aspects of an electrical circuit:

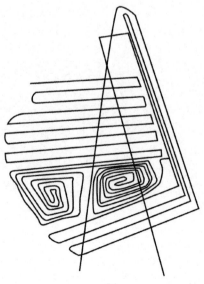

The middle part looks like two back-to-back bifilar coils, and above and below these coils are the zig-zag patterns. The coils and the zig-zag patterns are connected to each other by more lines which could represent wires, and the entire pattern, to me, looks like some type of inductor/resistor or inductive/capacitive resonant tank circuit, where energy is transferred back and forth between an inductor (coil) and a capacitor, building in intensity over time.

Ley Lines

There are also many lines at Nazca that go on straight as an arrow for miles, bringing to mind ley lines, or paths along the Earth's surface where, according to sensitives, energy flows. Where these ley lines meet, the energy swirls, and a concentration of energy occurs. Over time these spots became local holy sites, drawing humans to gather and experience, at least unconsciously, the effects of the increased energy. Even the Catholic Church encouraged communities to build churches on former pagan sites, both as a conscious effort to show the power of the church over pagan ways, and perhaps an unconscious acknowledgment of the inherent power of these sites. Wilhelm Reich postulated that any swirling pattern in nature was the result of two intersecting energy flows, from dust devils and tornados all the way up to solar systems and galaxies.

Human Ether Flows

As the Earth can be seen as a body with lines of energy traveling along its surface, humans can be seen as having bodies with lines of energy flowing on or just below their skin. But the etheric flows in and around the human body are both varied and complex. The main pulse of the human energy system is up and then down the spinal cord, somewhat like an alternating electric current in a conductor, first taking in the solar energy from above (prana), then absorbing the earth energy from below (kundalini). One flow induces the other, back and forth, throughout our entire life, and this main pulse induces other smaller flows, usually at right angles, which again induce still smaller flows, extending out towards and beyond the body's extremities.

It is tempting to view the body as a machine, complex indeed, but somehow understandable if we could only break down the overall system into all of its individual components, somewhat analogous to an old view of the solar system and the rest of the universe as some clockwork machine, controlled, or at least put in motion by a supreme being. Whatever or whoever is responsible for all existence created an organic universe, filled with a substance which binds and connects all things, and which is responsible for all life everywhere.

Acupuncture

The lines of energy, or meridians, near the skin's surface have been studied and mapped for thousands of years in Eastern medicine. Along these meridians are the acupuncture sites, which are low-resistance access points enabling specialists to unblock the energy flow and restore health to their patients. In general terms, the energy flows up along the inside of our legs, and down along the outside. It flows out of our fingers along the nail side and in along the palm side. Our back side has energy generally flowing down, and our front side up. There is more energy flowing out of our right hand and foot, and more into our left hand and foot. So we 'center' ourselves by sitting with our legs and arms touching, completing the circuit and building up or concentrating our body's energy. Controlling the bodily flow in this way can result in a general feeling of well-being and calm, fed by an increased amount of energy at hand.

The main goal of the art and science of acupuncture is to remove the blockages of the flow of the life-energy and restore the balance of this energy by inserting and stimulating tiny metal needles into the acupuncture sites on our skin. Modern acupuncture systems employ sensitive electronic equipment to aid the lay person in finding these sites throughout the human body.

Auras

So we draw in energy from the Earth below us, and breathe in energy from the air around us. The energy flows through us and emanates out of our bodies to produce the human aura, a constantly changing field of patterns of energy. This dynamic and ever-changing energy field consists of volumes of varying density levels and frequencies, nestled inside each other, and stretching toward the outside world. We sense the world through our aura, and communicate thoughts and emotions to the outside world through this field of energy.

A small number of people claim the ability to perceive this aura. Edgar Cayce, an American seer (known as the 'Sleeping Prophet' because of his trance-like states during readings) of the last century, claimed this ability, and was surprised to learn as a child that it was very rare indeed. Throughout his life he was able to catalog these perceptions, and learned to correlate the different colors perceived to specific states of mind, emotions, and personality types. His last written work, "Auras" (Cayce, 1945), explains the meanings of the different colors perceived based on his lifetime of experiences:

Red indicates vigor and energy, but as with all colors, depends on the specific shade. Pink represents immaturity, scarlet indicates a large ego, and dark red is associated with a high temper. All reds indicate nervous energy, and in their extremes show nervous turmoil.

Orange is the color of vitality, represented by the sun. A golden orange indicates self-control, while brownish orange in one's aura shows laziness or lack of purpose.

Golden <u>yellow</u> indicates health and well-being. People with a good amount of yellow in their auras are generally happy and friendly, and learn easily. A ruddy yellow could mean timidity or a weakness of will, or an inclination to be led by others.

<u>Green</u> is the color of healing, particularly if it has a dash of blue. Doctors and nurses invariably have green in their auras, but the green is frequently dominated by its neighboring colors. A lemony green shows deceit, while a deep healing green, even in small amounts, is positive.

<u>Blue</u> represents the spirit, and is the symbol of contemplation, prayer, and heaven. Pale blue indicates a struggle towards maturity, but at the same time progression. People with deep blue in their auras have

found their life's work and are immersed in it, though are apt to show moodiness.

Indigo and violet indicate seekers of all types, but as these people find what they are seeking, like a religion or a cause, the colors settle back into deep blue. Too much purple indicates an overbearing personality.

So the dark shades of the different colors generally denote more application, more will power, and more spirit, and the lighter shades shows less completion or maturity of the attributes associated with the individual colors. Food and environmental factors can affect the aura, which changes constantly to reflect our moods and states of mind, and evolves over our lifetime to show growth or retardation.

In the human aura, there are seven chakras (from Sanskrit for wheels or circles), from our base point to the top of our head, which are whirlpools of energy, each with their own frequency and color. Most sensitives see each chakra as two whirlpools of energy facing each other, swirling in opposite directions and transferring energy back and forth, somewhat analogous to an electrical transformer. The seven chakras:

	Element	Color	Gland Associations
Root	Earth	Red	Gonads Reproduction, Family Ties
Pelvic	Water	Orange	Spleen Creativity
Navel	Fire	Yellow	Adrenals Energy/Power, Confidence
Heart	Air	Green	Thymus Love, Compassion, Forgiveness
Throat	Ether	Blue	Thyroid Voice, Commun. of Emotion
Third Eye	Light	Indigo	Pituitary Intuition, ESP, Reg. of chakras
Crown	Cosmic Energy	Violet/White	Pineal Connection to Spiritual Realm

If we were able to see the human aura, we would see a constantly moving volume of energy made up of many layers and colors, interacting with both the environment and other living beings. We would see the healing energy of doctors and nurses flowing to their patients, the energy flowing from a devout crowd to a charismatic speaker, and the swirling interplay of the auras of humans in the act of making love. I like to believe that our future human evolution involves enhancement of this ability to see and experience the human aura.

As an illustration of this interplay of connected energy fields, I recall the birth of our second child. With my wife expecting at any time, we traveled to my sister-in-law's home about an hour away for Thanksgiving dinner. As soon as we arrived, my wife said that she was having contractions and we had to leave for home, where we had planned to give birth with the help of a doctor and a midwife. We hopped back in the car, and then I came to experience something that I had read about, but never quite believed – sympathetic contractions. Several times during this nervous ride home, I felt a sudden pain in my gut and let out a yelp of pain, followed immediately by my wife's yelp, signaling another contraction. In this extraordinary life experience, my wife and I were deeply connected by something, and a field of energy seems the most logical candidate. Action at a distance without a connecting medium has always been difficult to understand.

Rosicrucian Order

Throughout history, schools of thought have been based on the dual nature of the ever-present energy. The Rosicrucian Order (Order of the Rosy Cross), like many disciplines, base their healing exercises on the principle of balancing the body's positive and negative energy. They teach exercises that increase either positive or negative energy flow into specific areas of the body by stimulating ganglia, or networks of nerves. These ganglia then direct the influx of energy to an afflicted organ or body part to restore energy balance.

Rosicrucian breathing exercises include visualizing the energy flowing in with one's breath, and directing it to spin and concentrate itself in various parts of the body. Another exercise recommends walking with one's fingers curled. The energy (ether) passing through your fingers is thought to generate energy flow at right angles into the fingers and back into the body. The analogy that comes to mind is the basic electrical generator, where a bar magnet passing through a coil of wire produces a flow of electrons in the wire.

The theme of this chapter is the shapes or forms caused by the flowing ether. Imagine flying in a small plane across the landscape, looking at natural and man-made objects from a functional point of view. Trees are seen as natural energy sinks, drawing energy up from the Earth into the atmosphere. Water or oil tanks store these resources for future use, somewhat like batteries and capacitors store electrical charge in electric circuits. Rivers drain huge swaths of land, their meandering paths functioning to increase their efficiency. The rivers fill the oceans where water molecules evaporate and are borne by winds to re-condense over warm forests, repeating the cycle over and over.

Church Architecture

Buildings function to control flowing energy by their form, some better than others. As an example, the landscape is dotted with church steeples, said to draw worshippers' eyes and minds upward towards heaven. Steeples, or spires, function like other pointed objects in physics – they draw energy out from their points, making churches conduits of energy flowing up out of the ground, through the edifices, and out of their spires. Churches are built and rebuilt on the same terrestrial power points, using their structure's shapes to concentrate and direct the flow of the local energy, with worshippers reaping the benefits of immersing themselves in this flowing energy.

Russian church architecture uses the 'onion' shape atop many orthodox cathedrals, which has been shown to be a form that both stores energy and shoots it out of the point. The Chinese archetypal building with tiers of upturned roof panels is another variation on this theme. I actually thought for many years that the root of the word 'spire' had to come from the Latin word for breath or breathing, as in respire or conspire. Unfortunately, with a little research I learned that the root of 'spire' is the German word 'Spier', meaning 'tip of a blade of grass'. Oh well.

Feng Shui

All buildings and their surroundings can be viewed functionally in terms of energy flow. This is the ancient art of feng shui (literally wind-water). Where a dwelling sits in regard to streams, rises and falls in the landscape, and other features, determines the positive or negative flow of energy. As an example, the house that we lived in for many years had a small stream flowing about a hundred yards from the front door, parallel to its frontage, and there was a small bluff on either side of the front of the property. These were positive features for our house, the stream bringing flowing energy in front of us, and the bluffs providing protection like two parapets on a castle. A hill or mountain behind us it would also have been beneficial. You wouldn't want to live on the bend of a river, with the flow of the river's energy coming into your front door. According to the art of feng shui, that would cause energy to shoot right through your house for a negative effect. To combat this situation, mirrors or glass spheres would be hung in strategic spots, reflecting or dispersing the negative energy.

The principles of feng shui apply to the form of the building, to its placement in its environment, and to its interior design. Home placement rules include the avoidance of steep slopes, cemeteries, power stations, cluttered areas and roads that point into the house. Again, mountains or hills behind the house are auspicious. Because the front entrance is where the energy flows in, this area should be free of obstacles and well-lit, and should not be in a direct line with the back door, causing energy to simply sweep through the house. Likewise, staircases should not be facing the front door, as this would direct too much energy upstairs. Beds in bedrooms should not be placed facing the door or in front of a window, and you shouldn't sleep with your feet facing the door or under beams. Bathroom doors should always be closed, and before the toilet is flushed, its lid should be closed. There is much more to feng shui. In the many centuries of its use, different philosophical schools with their sets of tools have developed, and its use has grown and declined over time. Yet even in this modern enlightened age, the art is still used extensively in the East, and more so now in the West.

Dowsing

Feng shui is about flowing energy, and humans, at least unconsciously, feeling and responding to this energy. Another area of interest along these lines is the art of dowsing. The oldest known historical reference to dowsing is in a cave painting in Algeria, showing a tribesman holding two bent twigs searching for water, and is dated to be 8000 years old. Although dowsing has been practiced throughout history in scores of cultures, it is most often misunderstood, and its practitioners have been feared more than they've been respected. But when you need something as important as water, and someone can actually find that water for you, they tend to get your respect. Nowadays, water companies regularly employ dowsers, and local dowser associations can be contacted when needed. Drilling for water is expensive, so if a dowser can consistently point to productive spots, his or her worth is proven.

The theory behind dowsing is that flowing water creates an energy (etheric) field, and that humans can employ devices to help them feel this field. Practitioners tend to agree strongly that the human's mind needs to be 'tuned' to desire water. Indeed, as George Applegate, an author and researcher on the subject states, "But the desire to find water is simply the first step in raising the vibrations of our individual energy field. It needs to be reinforced by a firm belief that water can be found, and a faith in our ability to find it. Together these three components – desire, belief, and faith – step up the vibratory rate of our energy field, our dowsing consciousness, and in doing so program our subconscious mind that does the work for us." (Applegate, 2002)

Dowsing has been and continues to be controversial because the proof, one way or another, is at best illusive, and it is a system with a human at its center, who needs to be in a certain sustained state of mind for the system to be effective. Furthermore, like other phenomena of this kind, Earth energies are involved which vary by time of day, position of the Moon and Sun, and other factors.

Dr. Flanagan describes the ether as a nomenon, something incapable of being directly measured with today's tools, and everything else as phenomena, things measurable, even if what is really being measured is movement or a pattern of movement of the ether. As seen in the following chapter, some shapes or structures of specific dimensions, with distinct ether flow patterns, can exhibit special conditions inside the structure.

CHAPTER THREE: PYRAMIDS

Great Pyramid of Giza

The information presented in this chapter had its origins in a simple study done years ago, where I examined the reflection patterns inside the pyramid (or cone) shape, represented in 2 dimensions as an isosceles triangle. In particular, I wanted to see if there were something unique or unusual about the dimensions of the Great Pyramid at Giza, but I began by systematically examining the basic pyramid shape, with different base angles. As you will see, this study uncovered novel geometric relationships and patterns that tie together disciplines in both science and art.

If we see the Earth's energy fields as patterns of movement of the ether, then we have to look at the pyramid structure in its environment, in a sea of energy patterns that move in different directions, in, through and around the pyramid. The magnetic flow is precisely along the north-south edges of the pyramid, the rotation of the Earth (and the pyramid) is along the east-west parallel, and the flow of energy from the Earth to the atmosphere is straight up.

Traditional science teaches us that a pointed object shoots energy out of its point, and the sharper the point, the more energy movement through it. Lightning rods function on this principle; they draw electrons (negative) out of the Earth and shoot them out of the point, thus reducing the risk of a lightning strike by eliminating the concentration of negative energy. So I posit that one function of the pyramid must be the radiation of energy out of its top point. Is this substantiated by reputable reports of compasses in planes traveling directly over the pyramid's apex going haywire?

The assumption here is that energy is drawn out of the Earth and up through the pyramid. What happens to the energy inside the pyramid? There are only two possibilities: it either moves up from the base and drifts out of the side that it encounters, or it is reflected back inside. Because we know that the Great Pyramid was originally encased in a seven foot thick layer of smoothly polished alabaster, and because we know from Wilhelm Reich that this energy is reflected off of smooth surfaces at equal angles, I make another assumption, that most (maybe close to all) of the energy emanating from the Earth was meant to be reflected back inside the pyramid. The first and most obvious observation is that there is a concentration or build-up of energy inside the pyramid, which is partly the cause of 'pyramid effects', but not all, as will be seen. A four-sided pyramid can be graphically represented by an isosceles triangle, which has two sides of equal-length:

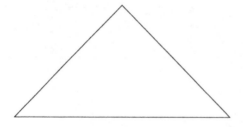

One line moving up represents one beam of energy, and there are innumerable beams of energy moving at any given time, then being reflected back inside:

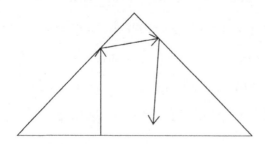

The 45° Pyramid

It becomes apparent after experimenting with triangles (pyramids) of different angles that there are certain specific triangles where the beams are reflected inside the shape a certain number of times, and shot back straight down to the base at 90°. The simplest one is the 45° pyramid:

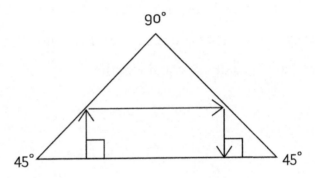

Standing Waves

The energy is reflected twice and directed down so that it either cuts the base at 90° or is again reflected straight back up along the same path. This is important because of what it may represent in the functioning of a pyramid. Using these assumptions, in a 45° pyramid all of the energy rises from the base and is bent across and down again perpendicular to the base, where it now encounters another beam rising up. In a space where two rays of energy with the same frequency and amplitude are traveling in exactly opposite directions, a <u>standing wave</u> is produced. This is an energetic state or pattern of movement where at certain points there is little or no excitation. These are called nodes. At other points in between there is great excitation, and these are called anti-nodes, or loops:

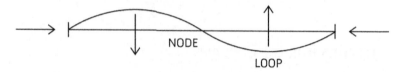

A standing wave appears to remain stationary, but in reality energy is being transferred in both directions. A resonance between the length of the vibrational period and the natural wavelength of the medium is established, and a stationary waveform becomes apparent. Is the standing wave part of the functioning of certain pyramids? What other pyramids produce this effect? The 45° pyramid is the middle point between the lowest (0°) and the highest (90°) possible pyramid. What happens inside of a 46° pyramid?

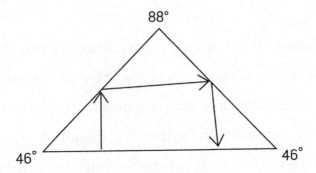

The energy is reflected twice and shot down not at 90°, but at 86°. There is no standing wave because the incoming and outgoing beams do not directly oppose each other. The 45° angle must be exact; a slight discrepancy destroys the standing wave effect.

Pyramids Greater Than 45°

If we imagine pulling up the apex, the next standing wave pyramid has a base of 60° pyramid, where there are three reflections: up, across to a right angle to the wall, and back down along the same line:

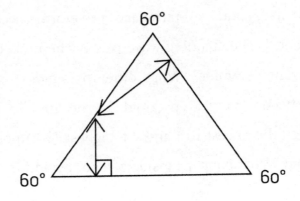

The 67.5° pyramid is next, with 4 reflections:

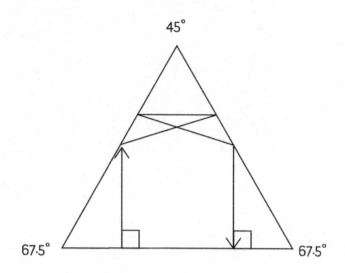

Next is the 72° pyramid: 75° pyramid:

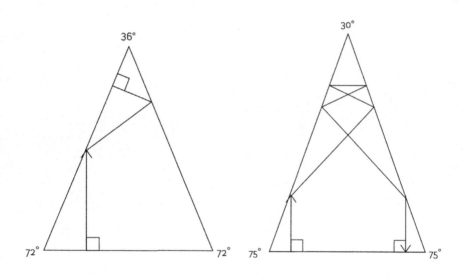

77.<u>142857</u>...° pyramid:

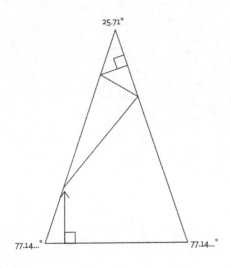

78.75° pyramid:

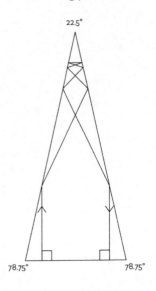

80° pyramid:

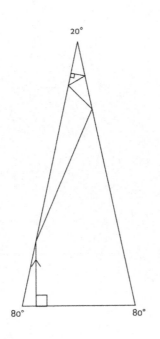

81° pyramid:

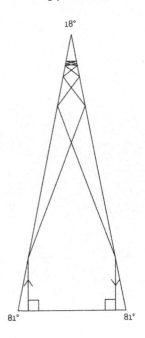

81.81**81**...° pyramid: 82.5° pyramid:

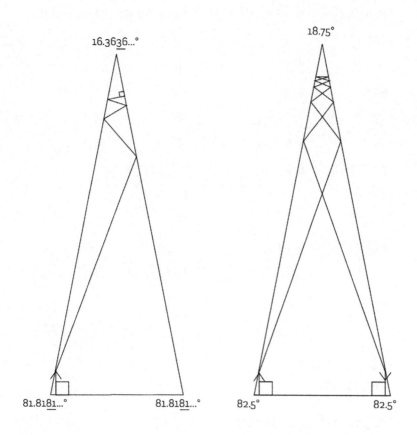

16.36**36**...°   18.75°

81.81**81**...°   81.81**81**...°   82.5°   82.5°

As the base angle gets steeper, the number of reflections keeps increasing, theoretically to infinity, because the apex angle is always approaching 90°, but never reaches it. On the following page is a list of the standing wave pyramids examined so far:

CHART I

Base Angle	Base Angle / 90°	Apex Angle	No. of Refl.
45°	1/2	90°	2
60°	2/3	60°	3
67.5°	3/4	45°	4
72°	4/5	36°	5
75°	5/6	30°	6
77.<u>142857</u>...°	6/7	25.<u>714285</u>...°	7
78.75°	7/8	22.5°	8
80°	8/9	20°	9
81°	9/10	18°	10
81.8<u>18</u>1...°	10/11	16.3<u>23</u>2...°	11
82.5°	11/12	15°	12
∨	∨	∨	∨

Equations

A general relationship is perceived for these standing wave phenomenon pyramids. The number of reflections (R) times the apex angle (A) equals 180°.

$$A \times R = 180°$$

The corollary equation from this in terms of the base and apex angles is that the number of reflections is equal to twice the base angle (B) divided by the apex angle, plus 1.

$$R = 2B/A + 1$$

Knowing from the patterns that the number of reflections must continue upwards in whole number integers, it is possible to use these equations to determine the rest of the series of standing wave pyramids. The 13-reflection pyramid must be 83.076925°.

$$Apex = 180° / 13 = 13.846153°$$

$$Base = 83.076925°$$

In the same way, the 14-reflection pyramid is 83.<u>571428</u>...°, etc.

Imagine a slim pyramid, or cone, which keeps getting sharper and sharper, with more and more beams of energy bouncing inside off of the sides. At certain discrete angles, and only at these angles, standing waves are produced inside the entire volume of the structure, and there is a heightened energetic state inside the structure. More energy inside means more energy shot out through the top point. More energy shot out of the top means more energy drawn in through the bottom of the structure.

This entire study is based on two key assumptions: there is some energy which emanates from the Earth, and this energy rises through the pyramid and at least some of it is reflected back inside off of the walls. From these assumptions it is possible, by simple geometry, to show that **there are certain base angles, and ranges of base angles, that set up conditions in a pyramid where some (all in many cases) of the beams, rays, or lines of energy rising through the floor are met by directly opposing beams of energy, producing standing waves and an increased state of energy in the interior of the structure.**

Theory of Partial Tones

In the first part of this chapter we looked at pyramids with base angles of 45° and greater. If we look again at this series of pyramids, we see an analogy to music, specifically the partial tone series in the study of musical acoustics. Partial tones, or harmonics, are small whole number multiples of the fundamental tone. When we hear, say, a cello sound a note, we don't really hear just one pitch, but a series of pitches which are mathematically connected, and which give that instrument its own characteristic sound.

The connection between this series of partial tones and the power angles in pyramids can be seen in the following chart. 180° divided by the apex angle gives the same proportional values as the partial tone's frequency divided by the frequency of the fundamental note. In a 3-dimensional structure, this value, 180° / Apex°, is equal to the number of reflections that it takes to bend the energy beam back 180°, producing a standing wave. When a string is plucked, basically a 2-dimensional vibration pattern is established (the string oscillating back and forth). This same proportional value, Fundamental Tone / Partial Tone, gives the number of loops in the vibrational pattern, as seen on the following page in Chart II:

Apex Angle Ratios	$\frac{180°/90°}{=}$ 2	$\frac{90°/60°}{=}$ 1.5	$\frac{60°/45°}{=}$ 1.333	$\frac{45°/36°}{=}$ 1.25	$\frac{36°/30°}{=}$ 1.2	$\frac{30°/25.7°}{=}$ 1.167	$\frac{25.7°/22.5°}{=}$ 1.143	
180° / Apex°	1	2	3	4	5	6	7	8
Apex Angle	180°	90°	60°	45°	36°	30°	25.71428°	22.5°
# Reflections	1	2	3	4	5	6	7	8
Standing Wave Pyramid Reflection Patterns — 3 Dimensions								
2 Dimensions								
Standing Wave String Oscillation Patterns								
# Reflections	1	2	3	4	5	6	7	8
Partial Tone	Fundamental C	1st Octave C_1	Fifth G_1	2nd Octave C_2	Major 3rd E_2	Fifth G_2	Minor 7th Bb_2	3rd Octave C_3
Partial / Fund	1	2	3	4	5	6	7	8
Partial Tone Ratios		$\frac{C_1/C}{=}$ 2	$\frac{G_1/C_1}{=}$ 1.5	$\frac{C_2/G_1}{=}$ 1.333	$\frac{E_2/C_2}{=}$ 1.25	$\frac{G_2/E_2}{=}$ 1.2	$\frac{Bb_2/G_2}{=}$ 1.167	$\frac{C_3/Bb_2}{=}$ 1.143

	22.5°/20°	20°/18°	18°/16.36°	16.36°/15°	15°/13.85°	13.85°/12.9°	12.9°/12°	12°/11.25°
	=	=	=	=	=	=	=	=
	1.125	1.111	1.1	1.091	1.079	1.078	1.075	1.067
...0°/Apex°	9	10	11	12	13	14	15	16
pex Angle	20°	18°	16.36°	15°	13.85°	12.85°	12°	11.25°
Reflections	9	10	11	12	13	14	15	16

Standing Wave Pyramid Reflection Patterns

Dimensions

Dimensions

Standing Wave String Oscillation Patterns

Reflections	9	10	11	12	13	14	15	16
artial Tone	Major 2nd	Major 3rd	Aug 4th	Fifth	Major 6th	Minor 7th	Major 7th	4th Octave
	D_3	E_3	~$F\#_3$	G_3	~A_3	Bb_3	B_3	C_4
artial / Fund	9	10	11	12	13	14	15	16
	D_3/C_3	E_3/D_3	$F\#_3/E_3$	$G_3/F\#_3$	A_3/G_3	Bb_3/A_3	B_3/Bb_3	C_4/B_3
	=	=	=	=	=	=	=	=
	1.125	1.111	1.1	1.091	1.079	1.078	1.075	1.067

This chart shows the analogy between the series of partial tones in the bottom half of the chart, and a specific series of 3-dimensional structures (pyramids or cones) in the top half. The partial tones, or harmonics, heard in any sound are shown as reflection patterns on a plucked string. When a string is plucked, many of these vibration patterns are occurring simultaneously, so that each sound heard is a combination, or overlapping, of different frequencies (notes). The proportional volumes of the different harmonic notes heard, along with its waveshape (attack, decay, sustain and release) give the sound its tone-quality, or timbre.

The series of partial tones, which is the foundation of the science of musical acoustics, has a direct analogy in the 3-dimensional world of structures seen in the top half of the chart. This series of pyramids or cones generates standing wave patterns, where energy is reflected inside the structure and shot straight back down to oppose incoming energy. In the bottom series, each time the number of reflections doubles, a new octave is reached, having twice the number of vibrations per second as the previous octave. In the top series, when the number of reflections doubles, the apex angle of the pyramid halves, showing a direct inverse analogy to the octave in music.

The two halves of the chart are the same series, set in different mediums. A new harmonic note is reached when a plucked string is divided evenly into its next number of oscillations. A new standing wave pyramid is reached when its height is increased in relation to its base so that the next number of reflections exactly sends the beam of energy in the opposite direction, creating a volume of standing waves of energy. In the bottom half, when the number of reflections doubles, the note is twice as high in pitch. In the top half, when the reflections double, the apex angles are halved, and the structure is twice as sharp.

5-Reflection Pyramids

There is another reflection pattern that is characterized by having 5 reflections, and using the floor for reflection. At about 48° base angle, beams of energy rising near the center of the base are reflected off of one side, then the other, then presumably, off of the floor, to be reflected again off both sides and down at 90° to the base. There are 5 reflections, and a standing wave is established, but not in the total volume of the pyramid, because beams rising too far from the center are not reflected 5 times and down. This 5-reflection pattern continues up to about a 54° base angle, where the opposite effect is produced. Beams rising too close to the center are not reflected 5 times and down perpendicular to the base.

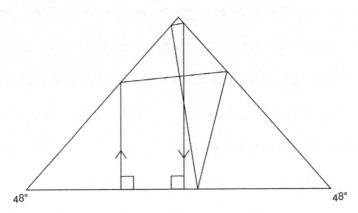

48°   48°

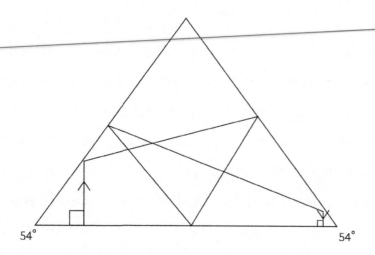

54°   54°

Throughout this entire range of base angle pyramids, 48° to 54°, the 5-reflection pattern persists, and it is one of the few cases where a standing wave pattern is generated not by a specific base angle, but inside a small range of base angles, each base angle with its associated total volume excited. It is also a pattern where there are reflections off of the floor as well as the walls. Because this pattern persists in a small range instead of one discrete base angle, the pattern can be said to be more resilient, or adaptive, to an imperfect structure.

With regard to this 5-reflection pattern, what is the base angle that produces the maximum interior volume excited by this standing wave pattern? The answer turns out to be the irrational number 5<u>1.42857</u>...°, where the last 6 digits repeat endlessly. A pyramid with a base angle of 5<u>1.42857</u>...° is the only one in this range (48° - 54°) where the <u>total</u> volume is excited by standing waves. Furthermore, the angles and their relationships to each other in this pyramid are unique in their symmetry and the way in which they complement each other. If the base angle is raised or lowered slightly, the symmetry is lost.

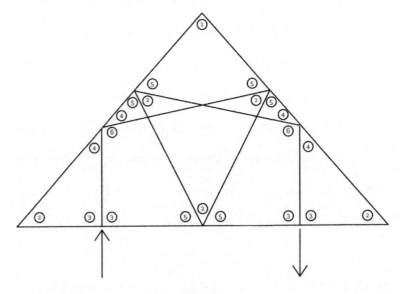

1) = 77.<u>142857</u>...°   2) = 51.<u>428571</u>...°   3) = 90°
4) = 38.<u>571428</u>...°   5) = 64.<u>285714</u>...°   6) = 102.<u>857142</u>...°

Every angle in this pyramid except 90° is an irrational number with the same 6 repeating digits, expressing a 7-ness in the angles (100 / 7 = 14.2857<u>142857</u>...). Furthermore, an analogy to musical pitches can be made in the 5<u>1.42857</u>...° pyramid. If we assign the note C to represent the base angle, then proportionally the rest of the angles have the following values in terms of musical pitch, forming a perfect C Seventh chord:

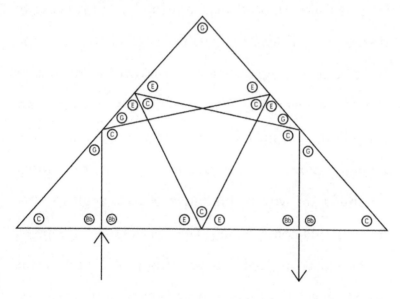

In astrology, the 'power angles' are harmonics, or equal divisions of a circle. 180°, 90°, 60°, 45°, etc. can all be multiplied by small whole numbers to get 360°.

If a circle could be stretched out like a string, the power angles would be the same harmonic divisions of the string that Pythagoras and the adepts used in their studies.

This range of base angles for the 5-reflection pattern definitely is real, at least geometrically if not energetically, and many pyramids were erected with a base angle at least close to the center of this range. Is it possible that there could be a surveying discrepancy of less than half a degree? What is the limit of surveying accuracy in a structure the size of the Great Pyramid? This 5-reflection pattern may provide insight into the knowledge of the builders of ancient pyramids all around the world. Most pyramids are within this range of base angles (48° – 54°). Is this coincidence? The Pyramid of Cheops is, at the most, only half a degree from the center of this range, where the total volume is excited. In any case, using these assumptions it is proposed here that the Great Pyramid's interior is almost entirely excited by standing waves. Did the ancient builders use this angle so that their imitators would at least excite most of their volume?

End Note

A guitarist lightly touches a point (node) on a guitar string, and then plucks the string. This node remains still, while the length of string between the node and the end of the string oscillates back and forth, producing a harmonic note. Points in a pyramid, following our two key assumptions, are still and calm compared to the area between them where the oscillation is maximum. A guitarist must exactly divide a string with his finger into a small whole number of portions to produce a harmonic note. In the same way, a pyramid must be of certain exact proportions to exhibit this standing wave phenomenon, when its base angle is a certain exact division of 90°.

CHAPTER FOUR: ASTROLOGY

There is a story that when Isaac Newton was being derided by Sir Edmond Halley for his interest in astrology, Newton's response was "Sir I have studied the matter, you have not!" This man who greatly advanced the sciences of optics, mathematics and physics also spent vast quantities of time studying biblical chronology, alchemy and astrology, which sound arcane to the modern thinker. But in Newton's time, science wasn't divided into many sub-disciplines, and a brilliant mind like his could seek out texts for self-study, come to understand the essence of each subject, and see the connections between each. The lesson here is to remain skeptical, yet keep your mind open.

Definition

Astrology has always been defined as the science of how the Sun, Moon and the planets affect life on Earth. It was once thought of as the mother-science, under whose umbrella all other studies flowed, but in modern times the subject has degenerated into a daily newspaper feature giving general advice for the population, grouped into twelve different astrological signs. This personal astrology (birth astrology) is a relatively new use of the system, becoming popular to the upper gentry in Europe a few centuries ago. Rulers could afford to have a court astrologer, but only a few of them were adept enough to predict anything of value to their sponsor. So let's take a look, skeptical but with an open mind, at this art/science to see if it can have some value to humans today.

Geocentric/Heliocentric

What is the system of astrology? There are really two basic schools – geocentric (Earth centered), and heliocentric (Sun centered). The argument for studying geocentric astrology is that we live on Earth, so we need to look out from our planet to find the forces affecting life here. Geocentric astrology includes the concept of astrological houses, which are times of day that correspond to the different aspects of life, and to some extent, the system of twelve zodiacal signs which are so familiar in modern culture. But the Earth is not the Sun, and the Sun is the proper center of the solar system, as well as the source of energy in the solar system. So in studying heliocentric astrology, we are trying to understand the conditions/weather of the whole system of star and planets. This work will consider only heliocentric astrology.

If we could see our solar system from far above the Sun (our North), we would see each planet rotating, and revolving in a counterclockwise direction around not the Sun, but the center of gravity of the solar system (CGSS). We would see that even the Sun is moved out from this common center of gravity, by as much as a solar diameter, when many planets pull on the Sun in the same general direction. Isaac Newton was said to pay great attention to this calculation, believing that the Sun's distance from the CGSS greatly affected the general weather of the solar system, as well as conditions on Earth.

Alignment/Declination

Each planet moves around the CGSS, 'connected' to it by gravity, and swimming in a sea of ether, radiating out from the Sun and filling the entire system. The angular position of a planet around the CGSS is the crucial measurement in this system, and as we will see, connects this planet to the other planets via 'power' angles. As an illustration of this principle, let me relate a childhood memory. I was the kid in the family (in the 50s and 60s) who was good at twisting and turning the old rabbit ears antenna on top of the TV for the best reception, and I noticed that the best reception usually corresponded with a certain angular position of the antenna, in respect to its vertical axis. But reception would also be good at regularly spaced angular intervals, as if the volume of space around the TV were divided into equal-spaced sections of an energy grid. Similarly, if I held the antenna as far from the TV as the wire would allow and walked around the TV, certain regularly-spaced angular positions would produce better reception. Is this an analog of how the solar system functions?

When two planets are in conjunction (angular alignment), they form a straight line to the Sun. But in order for the alignment to be exact, they must also have the same declination, or position above or below the plane of the ecliptic, which we define as the plane of the Earth's orbit around the Sun. Most planets revolve in a plane close to the plane of the ecliptic, but there is enough variation in the orbits of the planets to account for only rare exact planetary alignments. If two planets have different enough declinations, any effects caused by the planets' angular positions could be greatly attenuated.

The Planets

In the traditional astrological system, each planet has its own unique and innate characteristics. Mercury is associated with communication, travel and commerce. Venus rules love and lust. Mars is the god of war, and promotes strife and conflict. Jupiter is associated with luck and expansion, Saturn with wisdom, patience, hardships and fatalism. Uranus has been associated with electrical storms and sudden changes, and Neptune with psychic ability and uncertainty. Pluto, still considered a planet in astrology, is in charge of the subconscious. Because the outer planets travel slower, their alignments are longer lasting, creating more seasonal conditions. Alignments of the inner planets are more fleeting, with Mercury sometimes providing the triggering element in a solar system event, as our Moon triggers certain terrestrial events.

Let's say the Sun is at the center of this 360° circular grid, around which the planets revolve. The planets' orbits are elliptical, with the Sun (really the CGSS) at one focal point of each orbit. Each planet travels faster as it nears the Sun, and slows as it moves further away. Every planet also rotates in the same direction (counter-clockwise from above the solar system), with two notable exceptions. Uranus rotates in its orbit like a ball rolling on water, with its axis 90° different than the other planets. And Venus actually rotates backwards, something not known until the 1960s when the Russians radar mapped the planet. Also, this backwards rotation is slowing significantly; Venus' sidereal day increased by 6½ minutes in the 16 years between the Magellan and Venus Express spacecraft missions. This strongly suggests that Venus at some point was captured by the Sun's pull, rather than being formed with the rest of the solar system. As someone who has followed planetary news for the best part of my life, I consider this still a major solar system mystery.

As an interesting side note, before Venus' radar mapping, the Russian scientist Immanuel Velikovsky developed a complicated theory that the planet Venus began as a comet spewed from Jupiter's Great Red Spot, causing chaos in the solar system for millennia. In its travels, the comet Venus knocked Uranus on its side, causing its rolling rotation, and destroyed the planet that was between Mars and Jupiter, creating the asteroid belt. It then pushed the Earth and Mars out to their present orbital positions as (the planet) Venus settled into its new orbit. Velikovsky even said that these events, and many more related events, occurred in early human history, earning him an ouster from the Russian Academy of Science in the 1950s. Without giving too much credence to this elaborate theory, I note again that there has never been, to my knowledge, a good explanation for Venus' backward, and slowing, rotation.

Power Angles

When we say that two planets are 35° apart, we imagine lines radiating out from the Sun to each planet which form a 35° angle:

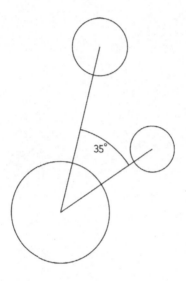

When certain angles connect two planets, the system of astrology says that something special happens. The power angles: 0° (alignment), 180° (opposition) and 90° are the most powerful angles. Generally speaking, multiples of 15° are the set of power angles in astrology. Furthermore, angles are grouped as hard or soft angles. Hard angles connecting planets are associated with the harsher characteristics of the planets, and include 0°, 180°, 90° and 45°. Soft angles include 120°, 60° and 30°, and bring out the more harmonious aspects of the coupled planets.

Archetypal Flow Patterns

YANG
Opposition
180°

YIN
Trine
120°

Cross
90°

60°

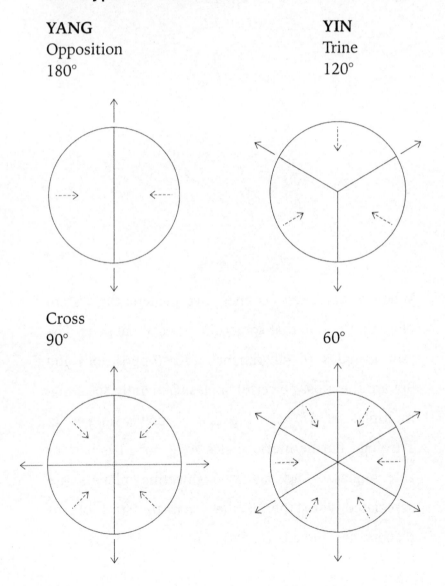

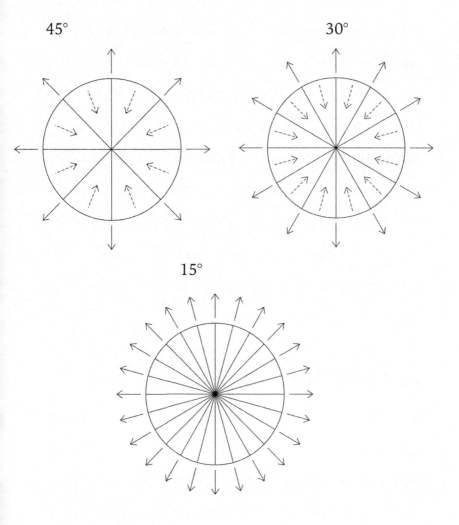

Divisions of a Circle

So much really comes down to how we divide a circle. We can divide the circle in half (180°), then divide these halves into quarters (90°), then divide these quarters into eighths (45°), like a pizza. These are the hard angles, bringing forth the masculine aspects of the connected planets. Or we can start by dividing the circle into thirds (120°), then dividing these thirds in half to make sixths (60°), then dividing these sixths in half to make twelfths (30°). These are the soft angles, causing the feminine qualities to dominate. Overall, the power angles are small whole number divisions of a circle: one-half, one-third, one-fourth, etc.

12°–15°–18° Series

The series of multiples of 15° is not the only series studied in astrology. Multiples of 18°, lesser known and not as strong in their effects, are sometimes used. So 18°, 36°, 54°, 72° and 90°, etc. would be the first part of the series, followed by 108°, 126°, 144°, 162° and 180°. Notice that 90° and 180° are common power angles between the two systems. I believe that it is possible that an additional series, multiples of 12°, could be used to further refine any predictions made. 12°, 24°, 36°, 48°, 60°, 72°, etc. would be the series. When we use the basic power angle, 15°, we are dividing the circle into 24 equal slices. The 18° series divides the circle into 20 parts, and the proposed 12° series divides it into 30 parts. If we stretch a circle out, we can see the relationships between the three systems:

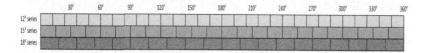

360 degrees is unity, unity is duality, and duality is symmetrical. The 12° series (multiples of 4°) contains halves, thirds and fifths of a circle. The 15° series (multiples of 5°) has halves, thirds, fourths and sixths, and the 18° series (multiples of 6°) has halves, fourths and fifths of a circle.

Connections

The typical view of the solar system is one where the Sun and planet-moon systems move mechanically through mostly empty space, affected only by gravity.

With an etheric viewpoint, our entire solar system is orbiting around the galactic center, with the Sun and planetary systems revolving and rotating (except for Venus) in a clockwise direction, but 'tacking' at about 31 degrees, relative to the galactic center. The solar wind, made of plasma, or atomic nuclei stripped of their electron shells, radiates out as far as we can measure, and the Earth's magnetic field shields us from this plasma, diverting most of it around our planet, and guiding the rest of it down into the atmosphere near the poles, creating the awesome aurorae. Additionally, the Sun's magnetic field interacts with each planet's magnetic field, connecting the Sun and the planets in a different etheric flow, similar to bar magnets hanging vertically in space. Also, the etheric flow out from the Sun follows along the line from the Sun to each planet, continuing past the planet. When two planets line up in conjunction, the flows out from the Sun combine into one stronger flow. If the planets are in exact alignment, where their declinations are also exactly the same, any effects from the conjunctions can be greatly enhanced.

Similarly, when planets are joined by the power angles described, something out of the ordinary can occur. Magnetic storms form on our Sun, sending plasma spewing throughout the entire solar system. The worst of these storms can disrupt electrical power grids and communication networks worldwide. Terrestrial magnetism and atmospheric pressure have been shown to be affected by solar weather, and in reality, all plant and animal life on Earth is constantly modulated by weather on the Sun, in turn caused, I believe, by certain planetary alignments.

What is it about these particular angles that is special? Is it something built into the fabric of space, or is it more like what was described in the last chapter, where it was shown that a volume can be filled with standing waves of energy when the structure is constructed of certain precise angles? If enough of the planets are closely aligned, the Sun can be drawn by them out from the solar system's center of gravity. When this happens, the Sun's gases, like everything else, are drawn toward this center of gravity, and can be ejected from the Sun's body at tremendous speed. Sunspots can form, usually in groups on one side of the Sun's equator.

John Nelson

What are the effects on Earth when planets are aligned in certain ways? Consider this example. After WWII, Radio Corporation of America, yes, RCA, charged an engineer, John Nelson, to find some way to predict radio weather for Earth - when radio signals would travel freer of electrical interference and have less static in their signal. RCA didn't care how he did it, but needed to be able to predict the basic quality of this 'weather', and to understand the system, if possible. John was an engineer, so he took an engineering approach. He didn't necessarily concern himself with ultimate causes, but focused on finding enough correlations to provide a path to accomplish his task.

There was some evidence that planetary alignments affected solar weather, so he waited for an effect, like a day when static dominated the airwaves, then made note of the associated planetary alignments that just occurred. After many events, patterns emerged that were tested over and over. The alignments were shown to precede the solar effects, providing evidence that they were the cause, and the changes in the Earth's radio weather the effect.

John Nelson rediscovered the basic laws of astrology. All of the power angles in planetary alignments were shown to affect solar weather, which affected the radio weather on Earth. Conjunction and opposition had the greatest effects, followed by 120°, 90°, 60°, etc. – all small whole number fractions of a circle, 360°. He found the hard angles negatively affected radio weather more than the soft angles, and discovered that the effect would greatly intensify if outer planets and inner planets were connected by the same power angles. John Nelson had no prejudice for or against the system of astrology. He didn't care. He just needed and found a way to accomplish his task. And of course precise positions of the planets are available many years in advance, so if the system is established, predictions could be made long in advance. He calculated his accuracy in predicting radio weather on Earth, once Pluto was included in the calculations, was about 91%.

Eclipses/Allais Effect

Now imagine our Moon revolving around us, completing its orbit in 27.3 days as observed from a particular point on Earth. When the Moon is close to being exactly behind the Earth, relative to the Sun, we call this full moon, because the full face of the Moon is lit by the Sun. And when the Moon is almost between the Earth and the Sun we call it new moon, or no moon, because the side of the Moon seen on

Earth is unlit. The Moon's approximate 4-week period peaks at full moon, when the Sun, Earth and Moon approximately line up, causing an increased etheric flow, which is harnessed by many plants and animals. The human emotional cycle seems to correlate with this lunar cycle, and there is a great deal of anecdotal evidence of strange human behavior at times of full moon. Rather than attribute weird human behavior to some mystical power of the Moon, my guess is that we simply have access to a larger pool of energy at these times, which we as humans too often use in odd, and sometimes harmful, ways.

Now once in a while the Moon passes exactly between the Earth and the Sun. This is called a solar eclipse because the Sun's face is eclipsed by the Moon. There have been reports of strange behavior exhibited by pendulums during solar eclipses. A pendulum is just a weight on a string which is set in motion, oscillating back and forth. Many science museums around the world feature these pendulums, the best of which have heavy weights and long cables to increase their stability. Over the course of one day, the pendulums' planes of oscillation appear to swing around in a circle. In truth, the pendulums' oscillating motion doesn't change direction; the Earth simply rotates beneath the pendulum.

During the solar eclipse of June 30, 1954, Maurice Allais, a French researcher with expertise in several fields, and who eventually won the Nobel Prize for Economics, reported a significant and unexplained anomaly in the movement of a Foucault pendulum. At the onset of the actual eclipse, the plane of oscillation of the pendulum shifted 13.5°, then moved back to its correct oscillating plane as soon as the eclipse was over. In the ensuing years, many researchers have repeated this experiment, some using very sophisticated equipment. Most researchers got similar anomalous results, while there were several notable exceptions. In the last few decades, the Allais Effect has not been seen in several experiments done. By my research, an unexplained phenomenon seemed to occur for a while, then stopped, until, I believe, interest in the anomaly faded.

Many theories have been proffered to explain this alleged anomaly, but none of them have satisfied the scientific community, especially considering that duplicate experiments produced conflicting results. If the effect is real, then we have a gap in our understanding of the basic workings of gravity, and/ or solar system functioning. But if it is real, we will eventually understand the process, and will have progressed in our knowledge. I tend to think of this anomaly as a function of the precision of the Sun-Moon-Earth alignment, where only the most exact conjunctions produce the largest results. Or perhaps there is a threshold of precision, which must be surpassed before the anomaly occurs. From an etheric framework, whatever effects would occur almost instantaneously, and would also be amplified by the number of bodies involved. Because the effect was seen in the 1950s through the 1970s (not always), and not seen over the last several decades could be attributed to some unknown environmental factor. Anomalies like this are fascinating, because if the effect is real it exists at the edge of human understanding, the place where science (knowledge) progresses.

11 Year Sunspot Cycle

Everything in the universe functions in cycles, and an intriguing one in our solar system is the approximate 11 year sunspot cycle. Sunspots, transient dark areas visible on the surface of the Sun, increase in number and peak, on average, every 11.1 years. These magnetic solar storms tend to appear in clumps, and can send magnetic flux lines and huge masses of plasma out to the far reaches of the solar system. Terrestrial magnetic activity and earthquake activity correlate to this sunspot cycle, and the human mass excitability index, which charts wars, migrations, and large human movements, usually violent, also corresponds to this pattern. The charts all show the same cycles, with peaks and troughs occurring at the same times.

Sunspot activity seems to correspond with some planetary alignments, and terrestrial magnetism peaks correspond to some planetary alignments. It seems Earth receives more energy at these planetary events, similar to what happens at full moon. These events are expressed in earthquakes, storms, and increased energy and activity in all life forms. Interestingly, each peak in sunspot activity is of a different polarity. The clumps of sunspots in one peak tend to occur mostly in the northern hemisphere of the Sun, while the next peak would have spots mainly in the southern hemisphere. Thus, the full cycle is about 22.2 years, or about 273 months, which also happens to be the length of the precipitation cycle on Earth.

My hope is that someone is intrigued enough to try to solve the mystery of this 11 year cycle. There are huge clues (Mars - Jupiter alignments, reversing polarity), and huge dead ends all around. I think it is mostly solvable, but very complex, and as much an art as it is a science. Planetary angular positions, declinations, momentum, position of the center of gravity of the solar system and other factors are involved, leaving the job to a good computer programmer with a background in math and a bent towards solving complex detective novels. Anyone?

Probability/Computers

Any subject with so many factors, which change every moment, needs to take advantage of modern computers and good software in order to advance to a reliable science someday, where predictions can be made and tested scientifically. Observe an event, and note the corresponding planetary positions. Check for repetition of similar planetary positions, and possible recurrence of the event. Make predictions, and test them. Get to the point where predictions can be expressed in terms of probability. And most of all, progress without prejudice, going where the data leads to advance the science of astrology.

CHAPTER FIVE: THE COMING ANTI-GRAVITY DEVICE

UFOs

The Theory

Biefeld-Brown Effect

Tank Circuit

Superconductivity

Resonance

Shapes

Characteristics

Lifters

UFOs

This chapter is an attempt to examine what I believe to be the next great technological leap for mankind - a true anti-gravity machine. For centuries humans have witnessed unidentified flying objects which seem capable of rising into the air effortlessly, and moving at incredible speeds. There are only so many possibilities as to what these craft really are. It's possible that the reports are all mistaken identifications - aircraft, weather balloons, cloud formations, etc., or faked sightings by people who crave attention. They could indeed be alien craft visiting Earth; we must at least count this as a possibility. Or some humans could already have developed this technology, whether aided by harvested alien technology or not. These are the possibilities.

Ancient texts and oral traditions of many cultures have described UFOs, so many that they cannot easily be discounted. Witnesses in recent history have included police and pilots, professionals who are trained to observe and report anomalous events accurately and dispassionately. And if airline pilots aren't impressive enough, astronauts describing lights and vehicles apparently watching them in space or on the Moon's surface take the cake. The thought that alien beings might have witnessed much of human progress is both frightening and inspiring. Even skeptical minds can be impressed by both the number of reports, and the details recounted. Mathematical arguments have been made that show the universe must contain advanced civilizations, some that would be capable of traveling great distances through space. Have we been visited by alien civilizations? I don't know, but my mind is open.

The Theory

It will be shown in this chapter that the Biefeld-Brown effect, discovered in 1921, was possibly the starting point of anti-gravity technology here, at least in the present civilization. If one looks at how far aeronautics, electronics, and other critical technologies have advanced since then, imagine where we could possibly be with anti-gravity technology. This work began with a story, which I attest to be true, of a demonstration on TV of a device which seemed to manipulate something in the room that knocked a man forcibly back into a wall. This technology, if real, could be used to move an object, not by applying a force on it, but by moving the substrate of matter, the ether, in a way as to cause the object to move.

But first, the theory. If gravity is a pull down from a large mass, it is hard to imagine a way to negate it, except by some sort of shielding, and we don't have a clue how to accomplish that. Even Einstein's concept of curved space doesn't seem to get us anywhere in this regard. To create a 'well' in the fabric of space to travel on, an object of enormous mass would be needed. However, if Patrick Flanagan's ideas are correct, a planet falls into its Sun because there is a greater concentration of the ether behind the planet compared to the side facing the Sun. Of course, this force is balanced by the planet's momentum, a function of its mass and velocity, which by itself would cause the planet to escape the Sun's 'grip'. If we extend this concept, then one way to move an object is to change the distribution of the ether around it. If we could move the ether from in front of an object to the area behind it, the object will move forward. It will be 'sucked' into the low density area in front of it, and 'pushed' by the high ether density in back.

It has been alleged in literally thousands of eyewitness reports of UFOs that these vehicles have the capability of making sharp turns at incredible velocities, or taking off and accelerating to great speeds in seconds. If there were some type of organic beings inside these crafts, then by everything we know both in physics and in physiology, these beings would be negatively affected by these maneuvers, literally having their bodies ripped apart by inertial forces. However, by moving the ether to propel the craft, it is thought that beings inside would not be affected, because the inertial system itself would be in motion, carrying the craft and crew along within it.

Biefeld-Brown Effect

To discuss the Biefeld-Brown effect, I will summarize the work of an author who goes by the name of Montalk. In an internet article where he references an old FATE magazine article, he explains that Thomas Townsend Brown worked with Dr. Biefeld at Denison University, in experiments that showed anomalous self-propulsive effects in high voltage capacitors. Their discovery was that a vacuum tube capacitor, basically two parallel metal plates suspended in a vacuum tube hanging in the air, would jump in the direction of the positively charged plate when a large charge was placed on the capacitor; that is, when electrons were moved to the negative plate and removed from the positive plate. So two horizontal, closely-spaced metal plates in a vacuum tube were charged quickly, the top (positive) plate lost its electrons and began to attract the ether from above the device, and the entire device momentarily lost enough of its weight to noticeably jump up.

Brown then progressed to experiment with two lead spheres connected by a nonconductive glass rod. One sphere was negatively charged, the other positively charged, with 120 thousand volts between them. When charged and suspended, the whole assembly arced toward the positive sphere, staying there as long as the charge remained. Brown then replaced the spheres with metal plates, separated by a strong dielectric material that could store a large amount of energy. Additional alternating layers were added to increase the effect. Brown called this a cellular gravitator, and was granted British patent #300,111. From the patent, it consisted of "numerous metal plates interleaved with dielectric plates, the entire block wrapped in insulating material and end plates connected to output electrodes and a spark gap to limit the input voltage. This device produced significant acceleration."

There were later reports about experiments performed in the 1940s with the US Navy, where Brown had saucer-shaped devices tethered to a pole which provided high voltage to the device. The 3-foot diameter disks had positive and negative electrodes on opposite sides, making them open-air high voltage capacitors. This configuration worked well both in air and in a vacuum, refuting the argument that it was simply an ion propulsion effect. After upgrading to 6-foot diameter disks, reports discontinued, allegedly becoming classified. In these devices, several factors were shown to affect the strength of the electrogravitational effect: voltage, current (some minimum is needed), strength of the dielectrical material, and overall capacitance of the device (increasing with the closeness of the metal plates, size of the plates, number of cells and increased asymmetry between the electrodes).

According to Montalk, the reason for the impulse in the direction of the positive plate is as follows: Simplify the plates to two single points, opposite in polarity and separated by a fixed distance. In addition to creating an electric field, this configuration generates a slight gravitational field, where the positive charge sucks in the surrounding ether, and the negative charge pushes out the ether. With the following geometry, the flow surrounding this dipole is biased in one direction. The positive charge sucks in from the left, and the negative charge pushes out toward the right, propelling the entire assembly in the direction of the positive charge, to the left. In etheric terms, decreased ether on the left side of the dipole pulls it to the left:

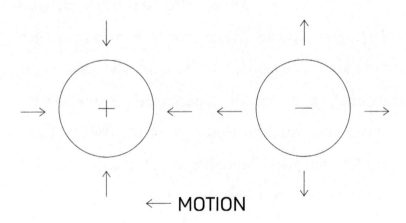

If the asymmetry is increased, specifically if the positive plate is made larger than the negative plate, the effect is enhanced. Brown also experimented with curved plates, one plate larger than the other:

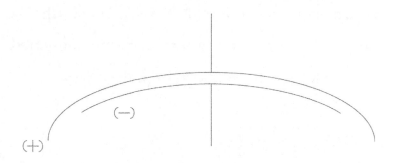

Remember that this basic phenomenon was discovered almost a century ago. It would be naïve to think governments or large corporations have not at least looked into this basic electrogravitic technology, considering the prize to be claimed. Thus my ease in believing that at least some of the UFOs could be of terrestrial origin. Assuming consistent secret government funding over a long time period, this technology could be tremendously advanced by now. If some alien technology has been shared or commandeered, and that's a huge if, then all bets are off as to where this technology could be now.

Tank Circuit

In these devices we've really only discussed a capacitor, basically two (or more) metal plates separated only by a strong dielectric material. Let's expand the circuit to include a coil, forming what is known as a tank circuit, where energy is transferred back and forth between the coil and the capacitor:

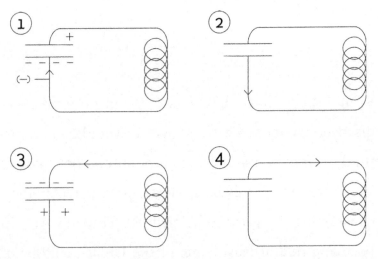

A stream of electrons is injected into this circuit and travels first to one plate of the capacitor (negative), building until the stream ceases. An opposite charge (positive) is simultaneously created on the other plate, and electrical energy builds up in the dielectric material between the plates. The electrons then leave the capacitor's negative plate and travel to the coil, basically creating an electromagnet, and storing magnetic energy in and around the coil. After the coil is fully charged, electrons continue on their path to the other plate of the capacitor, creating a negative plate here and storing energy in the dielectric material again. This oscillation continues back and forth, with both the capacitor and coil continually changing polarities. If electrons are injected only once into this circuit, then the oscillations continue until the natural resistive losses in the circuitry reduce the signal to zero.

What we call electrical energy stored in the dielectric of the capacitor, and magnetic energy stored in the coil, is in reality the ether, first compressed between the plates of the capacitor, then flowing around the coil in a magnetic flow. If the energy stored in the capacitor is equal to the energy in the coil, then it is called a resonant tank circuit, where both sides of this two-stroke electrical engine are balanced. The frequency of the oscillations where this occurs is the resonant frequency of this circuit.

Now if even a small signal is continually re-injected in the circuit at the same resonant frequency, and at the exact right time in its cycle, then a great deal of energy can quickly build up in the circuitry, so much so that a limiting device would need to be included to prevent the overloading of the components. The analogy of a child on a playground swing is appropriate here, where a small push given just after each peak causes great heights to be reached quickly. This continual small push to the child's swing results in a large amount of energy stored - kinetic energy when the swing is moving, and potential energy as the swing reaches its highest points. A small signal continually re-injected into our resonant tank circuit at the right moment in its cycle causes large amounts of energy (ether) to be built up and transferred back and forth between the capacitor and the coil.

Since the capacitor in this arrangement continually switches polarity, it would not suit the purpose of our simple antigravity device. Townsend Brown's disks used high-frequency pulses of direct current to achieve movement rather than an alternating signal. It is, however, easy to imagine modifications that would overcome this deficiency, using diodes, which are semiconductor devices that provide a one-way only path for electrons, to inhibit electrons where needed:

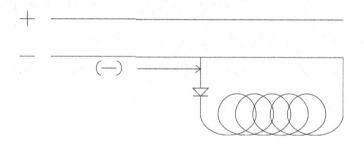

In the simplified circuit shown above, a pulse of electrons is injected just above the diode, which prevents the electrons from traveling down to the coil. The electrons are conducted to the center of the lower capacitor plate and amass there, making it negative in polarity. The top plate becomes positive in response. After the electrons fill the bottom plate, wires lead the electrons off of the edge of the plate to the coil, then through the one-way diode. Here the electrons meet another pulse and again fill the bottom plate, making it negative again, and the cycle repeats. From an etheric point of view, as the electric charges build on the capacitor plates, ether is drawn to the area between the plates, which is filled with a strong dielectric material to hold a large charge. Electrons then exit the negative plate and travel through the coil, creating a flow of ether through the coil as in a magnet. As the cycle repeats, the ether builds up and is more and more concentrated both in the capacitor and in and around the coil.

Superconductivity

Now consider the phenomenon of superconductivity. The resistive losses in wires and components, small as they are, can be reduced to zero or near zero by replacing the copper conductors with exotic alloys known as superconductors. These alloys become superconductive at very low temperatures, meaning there is <u>zero resistance</u> or impedance to the flow of electrons. The best of these alloys can attain a superconductive state with the use of liquid nitrogen for cooling, certainly a difficult technology to work with, but not an impossible one, while the goal is to find an alloy that superconducts at room temperature.

Superconductivity technology is used in sensitive devices like Magnetic Resonance Imagers (MRIs), which use coils of superconductive alloys to greatly increase their sensitivity. If our tank circuit were made with superconductive wire and components, and chilled with liquid nitrogen, the resistive losses would be cut to almost zero, keeping the energy oscillating back and between coil and capacitor almost forever, even with the smallest of pulses. For all we know, spaceships have been employing superconductivity for eons, even using the coldness of space to their great advantage.

Resonance

Now imagine this resonant tank circuit engine as an integral part of a vehicle whose natural resonant frequency matches the frequency of the tank circuit, or some simple harmonic (multiple) of this frequency. And imagine a small pulse of the correct frequency inputted to a circuit, quickly increasing the circuit's total energy until it can be transferred into the larger structures of the vehicle, which could easily feed a portion of the energy back to feed the tank circuit. Both the element and the whole would feed into and from each other, increasing the overall energy of the entire system – synergetic resonance.

Shapes

When it comes to the shape of an antigravity device, I believe that form has to follow function. If its function is to travel through space or a planet's atmosphere, then its basic form should enhance this movement, and even if space contains very few impediments to movement, some degree of streamlining would seem to be in order. Consider the classic flying saucer form, with a domed top:

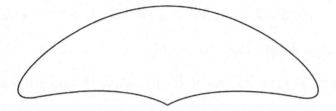

My assumption has been that the main movement mode of a saucer-shaped vehicle would be to go dome-first, at least in interplanetary space, with an additional possibility of a horizontal skimming type of motion for traveling in a planet's atmosphere. A saucer could also naturally house some type of Biefeld-Brown engine, pulling in the ether from above the vehicle, and directing it out below. If some type of tank circuit is employed, perhaps even the coil could assist the ether flow, with both parts contributing to the functioning of the device – moving through space. Putting it all together:

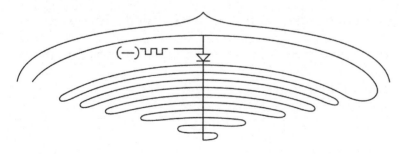

Characteristics

Let's review the main characteristics, or features, of our hypothetical antigravity vehicle: It has an engine with a large capacitor and coil, forming a circuit where energy oscillates back and forth and increases in power with the input of a signal of the correct frequency. A high voltage source would be necessary, with some, not a lot, of current, and a strong dielectric material would be necessary to ensure a high capacitance value. Reich's Orgonomic layering might be used to concentrate the energy, and his cloudbuster concept might provide further help in moving the ether. The shape should fit the function of the device, even considering the thin density of space. The pulsation of the input signal should be resonant with natural frequency of the engine, which should resonate with the shape and dimensions of the whole structure. Superconductivity would ensure super efficiency.

Lifters

In recent years hobbyists and experimenters have produced what are called lifters, very light-weight machines that demonstrate, to some extent, an anti-gravity effect. They use high voltage, high frequency power supplies, and can be seen in numerous internet videos. The devices are tethered by wires that supply power; none that have been shown can lift the weight of their own power supplies. Indeed, if they weren't made of lightweight materials like balsa wood, they wouldn't be able to lift off the ground, even without their power supplies. Most are designed to produce an ion wind, exhibiting motion by a charge of ions inducing air flow. Only a few are designed to downplay the production of this ionic wind and emphasize the purely electrogravitic aspect of the device.

If these electrogravitic, sometimes called electrokinetic, designs were improved to emphasize elements like a more efficient shape, resonance, and possible layering like Reich's Orgone Accumulator, then lifters' capabilities might progress to a new level. The public could witness their evolution, hopefully using the openness of the internet to bypass any impediments thrown in by say, a government that might not want us to know about this technology.

In reality, the goal is to come up with a new form of propulsion, by manipulating the ether to produce movement of an object. Using this device to escape the gravitational hold of a planet would be its ultimate test, and it would theoretically be able to continue into interplanetary space, since the ether fills all of space. In my mind, this new (to us) technology is the next huge breakthrough that must be developed for mankind to progress forward. In the end though, I'm convinced that we are just re-inventing technology (publicly at least), that has been re-invented countless times, as many civilizations have done at some point in their evolution. How advanced were these civilizations? We'll probably never know, but it would be foolish to assume that ours is the most advanced, and it would be sheer hubris to believe that in the incredible vastness of more than 200 billion galaxies there have not been civilizations so advanced that their science would seem like pure magic to us here in this age.

LAST WORD

My goal in writing this is to present the reader with a different way of understanding nature. Essentially, everything is made of the ether, and everything is interconnected by the ether. All that we see and feel is really an infinite variety of complex flow patterns of the ether, swirling and combining with other flow patterns to fill the universe. Additionally, our solar system seems to have an intrinsic geometric field, where specific angular relationships of its planets produce energetic states. And this entire series of mathematical relationships is not only expressed in the space of our solar system, but in buildings on Earth, and even in the science of musical acoustics.

I have no doubt that much of what I've presented is at best incomplete, and at worst woefully wrong. That's ok. I'm much more afraid of saying nothing than I am of saying something incorrect. My hope is that minds are opened to new possibilities. If a topic interests you, do your own research. Information has never been more accessible, but care must be taken to ensure that your sources are reliable. Read all that you can from multiple sources and make up your own mind, and be willing to adapt your views when better information emerges.

On a personal note, I'm not sure this would have been completed had I not contracted a somewhat debilitating ailment. It's given me time to think about my life's goals, and assemble the thoughts of a lifetime into this work. I like to think that I've learned to see the beauty in all of life, even in the tragedies that each of us endure. So my advice - don't curse your fate, but embrace the new possibilities that are put in front of you every day. As they say, when life gives you lemons, squeeze them for all it's worth, sit back, and enjoy a tall cool glass of lemonade.

Daniel N. Spatucci

References:

Reich, W., <u>Ether, God and Devil/Cosmic Superimposition</u>, 1973

Reichenbach, K., <u>The Odic Force</u>, 1852

Flanagan, G.P., <u>Pyramid Power</u>, 1975

Cayce. E., <u>Auras</u>, 1945

Applegate, G., <u>The Complete Book of Dowsing</u>, 2002

Background:

Goodavage, J., <u>Astrology: The Space Age Science</u>, 1966

ABOUT THE AUTHOR

Daniel N. Spatucci was employed for almost three decades in various technical positions at a major Defense contractor, producing advanced radar systems for the U.S. military and its allies. Daniel has also enjoyed a lifelong interest in music, singing and playing in choruses and bands across the country. In 2018, he and his wife Lisa retired to Florida, where his love for music and science continue to fill his time.

Learn more at **www.etherflowsbook.com**

Printed in Great Britain
by Amazon

26454269R00096